7th Symposium on Large TPCs for Low Energy Rare Event Detection 2014

Journal of Physics: Conference Series Volume 650

Paris, France
15 - 17 December 2014

ISBN: 978-1-5108-1813-2
ISSN: 1742-6588

Printed from e-media with permission by:

Curran Associates, Inc.
57 Morehouse Lane
Red Hook, NY 12571

proceedings.com

Some format issues inherent in the e-media version may also appear in this print version.

Printed by Curran Associates, Inc. (2016)

For permission requests, please contact the Institute of Physics
at the address below.

Institute of Physics
Dirac House, Temple Back
Bristol BS1 6BE UK

Phone: 44 1 17 929 7481
Fax: 44 1 17 920 0979

techtracking@iop.org

Additional copies of this publication are available from:

Curran Associates, Inc.
57 Morehouse Lane
Red Hook, NY 12571 USA
Phone: 845-758-0400
Fax: 845-758-2634
Email: curran@proceedings.com
Web: www.proceedings.com

Table of contents

Volume 650

7th International Symposium on Large TPCs for Low-Energy Rare Event Detection 15–17 December 2014, Paris, France

Accepted papers received: 12 October 2015
Published online: 16 November 2015

Preface

Papers

Preface

The seventh "International Symposium on Large TPCs for Low-Energy Rare Event Detection", took place in Paris between the 15th and 17th of December 2014 at the Institute of Astroparticle Physics (APC) campus – Paris Diderot University.
As usual the conference was organized during the week before Christmas, which seems to be convenient for most of the people and occurs every two years with almost 120 participants attending.
Many people contributed to the success of the conference, but the organizers would particularly like to thank the management of APC for providing the nice Buffon auditorium and infrastructure. We also acknowledge the valuable support of DSM-Irfu and the University of Zaragoza.
The scientific program consisted of plenary sessions including the following topics with theoretical and experimental lectures:

- Low energy neutrino physics
- Neutrinoless double beta decay process
- Dark matter searches
- Axion and especially solar axion searches
- Space experiments and gamma-ray polarimetry
- New detector R&D and future experiments

The connection between dark matter searches and the LHC accelerator results was pointed out in several theoretical presentations (G. Servant, O. Buchmueller), that were complemented by a review on dark matter experiments (P. Di Stefano) and the presentation of several experimental results on WIMP searches, using gaseous or liquid TPCs (xenon or argon). Very-low energy threshold detection for ultra-light dark matter searches, using the novel spherical TPC (I. Giomataris) or TPCs Micromegas detector read-out with directional recoil capability (MIMAC, NEWAGE) have been also presented.
A. Lindner gave a summary on searches for WISPs and the necessity of very low flux detectors, while several other talks focused on current solar axion searches using Micromegas detectors (J. G. Garza), Ingrid detectors (K. Desch), the future IAXO helioscope (E. Ferrer-Ribas) and the use of the Spherical detectors for searching KK axions at LSM (J. Galan).
The use of large liquid argon TPCs or specialized gaseous TPCs for low energy neutrino physics was pointed out in several presentations for current investigations (T2K experiment) or future projects on neutrino oscillations – CP violation studies in the next high intensity neutrino facilities at FNAL or in Japan.
Technical detector R&D and especially Hole Multipliers for use on double phase liquid argon for neutrino detection or liquid xenon for dark matter searches were reviewed in several interesting talks.
F. Piquemal presented a review on neutrino-less double-beta decay (DBD) searches. The results from EXO200 and the plans for NEXO and Barium tagging were reviewed (D. Sinclair) along with new ideas and recent developments regarding high pressure xenon-136 TPC (Y. Nakajima, D. González-Diaz) or selenium hexafluoride as a candidate material for neutrino-less DBD searches (D. Nygren) .
Alvaro De Rujula gave an inspired summary talk with an emphasis on the past, present and future of particle physics.

Journal of Physics: Conference Series **650** (2015) 011001 doi:10.1088/1742-6596/650/1/011001

Finally, we would like to thank the secretaries (Martine Oger, Angèle Sénée and Sarodia Vydelingum) and computing engineer Nicolas Pailley, as well as David Attié for their efficient help in the organization.

P. Colas
I. Giomataris
I. Irastorza
Th. Patzak
Conference organizers

Peer review statement

All papers published in this volume of *Journal of Physics: Conference Series* have been peer reviewed through processes administered by the proceedings Editors. Reviews were conducted by expert referees to the professional and scientific standards expected of a proceedings journal published by IOP Publishing.

Journal of Physics: Conference Series **650** (2015) 012001 doi:10.1088/1742-6596/650/1/012001

AMS-02 in Space: Physics Results

Nicola Tomassetti, on behalf of the AMS collaboration
LPSC, Université Grenoble – Alpes, CNRS/IN2P3 – Grenoble, France
E-mail: nicola.tomassetti@lpsc.in2p3.fr

Abstract. The Alpha Magnetic Spectrometer (AMS-02) is a particle physics experiment designed to study origin and nature of Galactic Cosmic Rays (CRs) up to TeV energies from space. With its high sensitivity, long exposure and excellent identification capabilities, AMS is conducting a unique mission of fundamental physics research in space. To date, more than 60 billion CR events have been collected by AMS. The new results on CR leptons and the analysis and light-nuclei are presented and discussed. The new leptonic data indicate the existence of new sources of high-energy CR leptons, that may arise either by dark-matter particles annihilation or by nearby astrophysical sources of $e^{\pm}$ pairs. Future data at higher energies and forthcoming measurements on the antiproton spectrum and the boron-to-carbon ratio will be crucial in providing the discrimination among the different scenario.

1. Introduction

The Alpha Magnetic Spectrometer (AMS) is an international project devoted to search for new physics phenomena, such as primordial antimatter or dark matter, by means of direct detection of high-energy cosmic-ray (CR) particles in space. The final version of the detector, AMS-02, was successfully installed and activated on the International Space Station on May 19th 2011. AMS-02 is capable of measuring CR leptons and nuclei, from hydrogen up to iron, from hundreds MeV up to $\sim$1 TeV of energy, with unprecedent precision and sensitivity. The AMS-02 data are expected to significantly improve our understanding of the CR acceleration and propagation processes in the Galaxy. The status of AMS measurements of CR leptons and light-nuclei is reviewed in these proceedings. The layout of the AMS-02 instrument and its complete description can be found in earlier papers [1]. The detector is composed of a Silicon Tracker, a Transition Radiation detector (TRD), a Time of Flight system (TOF), a permanent magnet, an array of AntiCoincidence Counters (ACC), a Ring imaging Cherenkov detector (RICH), and an Electromagnetic Calorimeter (ECAL). Redundant measurements ensure a complete particle identification and allow for a careful study of interactions inside the detector [2, 3]. The TRD [4] allows for lepton/hadron separation. It is made of 5248 proportional chambers, filled with xenon and carbon-dioxide, arranged in 20 layers of fiber–fleece radiators. Four TOF planes of scintillators [5], located above and below the magnet case, allow for velocity measurements, charge measurements, and trigger. The Tracker is build of nine layers of silicon micro-strip detectors [6], from L1 to L9, distributed over the instrument. Layers from L2 to L8 are assembled with the permanent magnet, that has a magnetic field strength of 0.15 T. L1 is located on top of the TRD, and L9 is located between between RICH and ECAL. The Tracker measures the rigidity R up to $R \approx 2$ TV ($R \equiv p/Z$, momentum to charge ratio) for charge one particles, and charge measurement up to $Z = 28$. The RICH [7] is made of a radiator layer, a conical mirror,

Journal of Physics: Conference Series **650** (2015) 012001 doi:10.1088/1742-6596/650/1/012001

and a photomultiplier plane which detects Cherenkov light. Two radiators, with NaF and aerogel (refractive indices n=1.34 and n=1.04), are used to provide high-precision measurement of the particle velocities and charge. The ECAL [8] is made of lead and scintillating fibers arranged in 16 layers, providing 17 radiation lengths of detecting medium. This allows for a precise reconstruction of the energy up to several TeV. A boosted decision tree classifier (BDT) is used to identify leptons based on the topology of their showers. Onboard the ISS, AMS-02 is orbiting the Earth at an altitude of about 400 km and with inclination of 51.6°. The average trigger rate is about 600 Hz, the average event size is 2 kByte. The minimal down-link bandwidth is 9 Mbit/s. AMS-02 collects about 1.5 billion CRs during each month of operation. The detector is controlled from the AMS-02 Payload Operations Control Center (POCC) at CERN, Geneva. Intensive time-dependent calibrations have been performed to all sub-detectors. No significant degradation of the sub-systems has been observed during 3 years of operation in the ISS.

2. Results on cosmic-ray leptons

The new results on CR leptons are based on the data collected during the initial 3 years of operations on the ISS (May 2011 – May 2014), which is about 16% of the total expected data sample. In the measurement of leptons, the TOF is used to select relativistic particles traversing AMS-02 in the downward direction. The signals in TRD and ECAL are used to discriminate the leptonic component from the hadronic background. A track reconstructed from L2 to L8 and matching the TRD and ECAL signals is used to select clean $Z = 1$ particles. For the

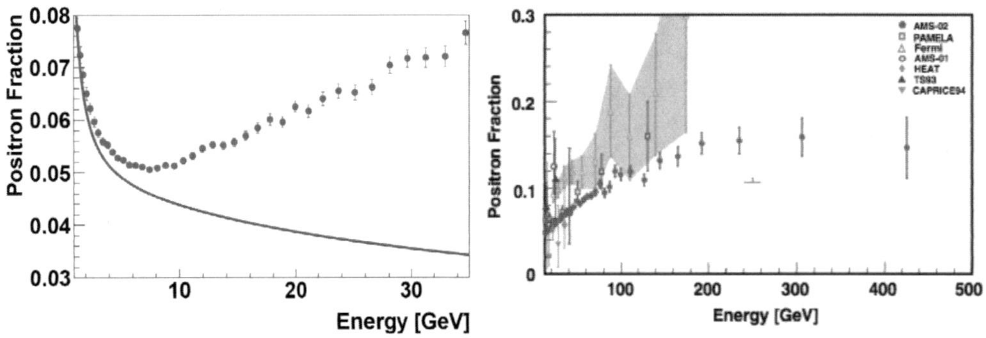

Figure 1. AMS-02 positron fraction data from 1 to 35 GeV (left) and from 10 to 500 GeV (right) compared with previous measurements.

lepton/hadron separation, three methods are used: a TRD-likelihood estimator, an ECAL-BDT estimator, and the E/R ratio by combining ECAL and Tracker. This redundancy allowed to study the performance of each discrimination method with the data. Various analyses with different combinations of cuts or template fits were also tested as a cross-check. An important source of background comes from *charge-confusion* of electron events, arising from the finite rigidity resolution of the Tracker, or from the emission of secondary particles. The charge-confusion contribution is studied from data and Monte Carlo simulations [1]. A high-energy measurement of the positron fraction (from 1 to 500 GeV) has been recently reported [9]. The results are shown in Fig.1. The left panel shows the fraction up to 35 GeV of energy compared with the standard conventional calculations where CR electrons are emitted from supernova remnants (SNRs) and secondary $e^{\pm}$ arise from collisions of CR nuclei with the interstellar matter (ISM). This model predicts that the positron fraction decreases with energy. Above 10 GeV, the AMS data show a persistent rise up to 200 GeV, in clear contrast with the conventional picture, followed by an intriguing flattening at higher energies (right panel).

3. Results on cosmic-ray nuclei

Light nuclei with $Z > 3$ constitute $\sim 1\%$ of the total CR flux. Rare elements such as ^{2}H, ^{3}He

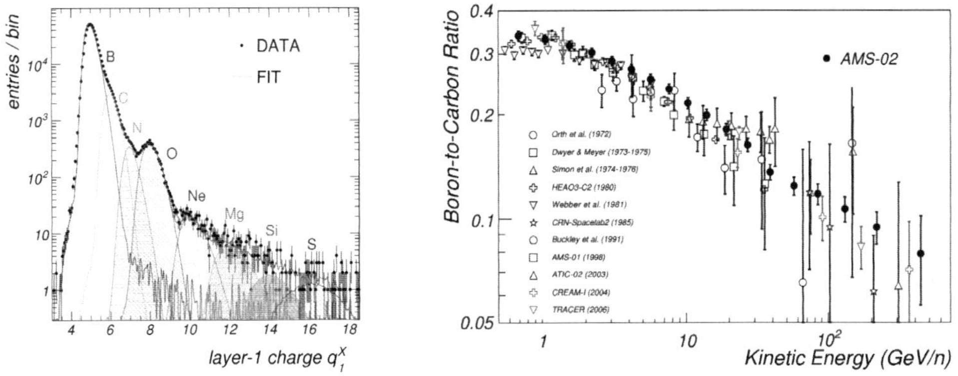

Figure 2. Left: Layer-1 Charge distribution for a loose Boron selection on TOF and Tracker (L2-L8). Fragmentation from heavier nuclei interacting in material below L1, such as C→B or O→B, can be estimated by a fit. Right: the measured B/C ratio from 0.5 to 700 GeV/n [10].

and Li-Be-B are believed to be of secondary origin, i.e. produced by collisions primary CRs (p, ^{4}He, C-N-O, or Fe) with the gas nuclei of the ISM. Secondary to primary ratios such as ^{3}He/^{4}He, B/C, or F/Ne are used to discriminate among astrophysical models of CR propagation in the Galaxy [11]. Furthermore, the B/C ratio at $\gtrsim 100$ GeV/n represents a crucial tool to understand open problems in CR physics such as the observed structures in primary CR spectra [12] or possible SNR origin for the positron excess [13, 14, 15]. The measurement of the B/C ratio is a prime physics goal of AMS-02. The analysis of the first 24 months of data has been presented in [10]. With AMS-02, high purity samples of B and C events have been obtained using the TOF and Tracker charge estimators, with a charge selection efficiency of 98% and a charge mis-identification probability less than 10^{-4}. Monte-Carlo simulated events are produced using GEANT4-4.9.4 package and DPMJET-II.5 [16, 17], including a description of detector geometry, electromagnetic and hadronic physics processes, as well as the full digitization chain in order to reproduce the signals of all sub-detector. The several charge detectors above and below the inner core of the spectrometer have been used to improve the overall charge separation capability as well as to select clean data samples of non-interacting CR particles. In particular, they allows for a data-driven characterization of nuclear interactions occurring at different levels in the spectrometers. Data and Monte-Carlo show an agreement at $\mathcal{O}(\%)$ level. Figure 2 (left) shows the L1 charge distribution obtained after selecting pure B samples in Inner-Tracker (L2-L8) and TOF. The colored curves are the single-layer charge responses for known elements. The data (black markers) are described by a combination of these templates using Boron (94%), Carbon (4%), and heavier elements ($\sim 1\%$). These numbers represent the probability for top-of-instrument processes of charge-changing fragmentation, obtained with flight data. In this figure, no cuts are applied to the TRD signal. Simple cuts on the TRD or L1 charge allows to highly-efficiently suppress these events. The measured B/C ratio is shown in Fig. 2 (right) as a function of the kinetic energy obtained from the Tracker rigidity. Effect of bin-to-bin energy migration due to the finite resolution of the instrument has been accounted using an unfolding procedure. The ratio decreases smoothly with energy. The B/C behavior at TeV energies will be become more transpaent with more data.

Journal of Physics: Conference Series **650** (2015) 012001 doi:10.1088/1742-6596/650/1/012001

4. Conclusions and discussion

New AMS results on the positron fraction show persistent rise up at $\sim 10-200$ GeV of energy, in clear contrast with conventional model predictions, followed by an intriguing flattening at higher energies. From the measurement of the single spectra of e^+ and e^- [18], the data show that both components e^- and e^+ harden smoothly at $\gtrsim 30$ GeV. These data are also consistent with the measured "all-electron" spectrum, $(e^- + e^+)$, which has been independently determined up to $E = 1$ TeV [19]. All these data indicate the existence of a new high-energy source of e^+ and e^-, which can be interpreted either in terms of dark matter (DM) particles annihilation or in terms of nearby astrophysical sources of $e^\pm$. The DM interpretation require annihilating particles with mass of the order of ~ 1 TeV, and may predict possible signals in the $\bar{p}/p$ ratio, depending on the specific DM–DM annihilation channels [20, 21]. The existing antiproton data show no clear excess within the uncertainties in the data and in the model predictions. In the second class, a recent work demonstrated that the AMS-02 data are described well in terms of $e^\pm$ pairs emission from pulsar wind nebulae (PWNe) [22]. The PWN scenario gives no signatures in the antiproton channel. It was also proposed that high-energy $e^\pm$ can be produced inside old-SNRs via interactions of CR protons with the background medium [23, 13]. This mechanism predicts remarkable signatures in both the $\bar{p}/p$ ratio [24] and the B/C ratio [13], that may be detectable by AMS-02 [14]. The data show no prominent signatures in the B/C ratio [10, 15]. The present models are affected by large astrophysical uncertainties that may be dramatically reduced with new data on CR nuclei [25, 26].

5. Acknowledgement

This work is supported by acknowledged persons and institutions in [1] and by the LabEx grant ENIGMASS.

[1] Aguilar, M., et alii, 2013, PRL, 110, 141102
[2] Tomassetti, N., & Oliva, A., 2013, 33th ICRC, 896, Rio de Janeiro
[3] Saouter, P., et alii, 2013, 33th ICRC, 789, Rio de Janeiro
[4] Kirn, T., et alii, 2013, NIM, A 706, 43
[5] Basili, A., et alii, 2013, NIM, A 707, 99
[6] Alpat, B., et alii, 2010, NIM, A 613, 207
[7] Aguilar, M., et alii, 2010, NIM, A 614, 237
[8] Adloff, C., et alii, 2013, NIM, A 714, 147
[9] Accardo, L., et alii, 2014, PRL, 113, 121101
[10] Oliva, A., 2013, 33th ICRC, 1266, Rio de Janeiro
[11] Strong, A. W., et alii, 2007, Ann.Rev.Nucl.&Part.Sci., 57, 285–327
[12] Tomassetti, N., 2012, ApJ, 752, L13
[13] Mertsch, P., & Sarkar, S., 2014, PRD, 90, 061301
[14] Tomassetti, N., & Donato, F., 2012, A&A, 544, A16
[15] Tomassetti, N., & Donato, F., 2015, astro-ph/1502.06150
[16] Agostinelli, S., et alii, 2003, NIM A, 506 3, 250–303
[17] Ranft, J., 1995, PRD, 51, 64
[18] Aguilar, M., et alii,2014, PRL, 113, 121102
[19] Aguilar, M., et alii, 2014, PRL, 113, 221102
[20] Boudad, M., 2014, arXiv:1410.3799
[21] Geng, C. Q., et alii, 2014, arXiv:1411.4450
[22] Di Mauro, et alii, 2014, JCAP, 1404, 006
[23] Blasi, P., 2009, PRL, 103, 051104
[24] Blasi, P., & Serpico, P. D., 2009, PRL, 103, 081013
[25] Tomassetti, N., 2012, Astrophys.Space Sci., 342, 131-136
[26] Maurin, D., et alii, 2010, A&A, 516, A67

Journal of Physics: Conference Series **650** (2015) 012002 doi:10.1088/1742-6596/650/1/012002

Detecting the barium daughter in ^{136}Xe 0-νββ decay using single-molecule fluorescence imaging techniques

David R. Nygren

Department of Physics, University of Texas at Arlington
Box 19059, Arlington, TX 76015

nygren@uta.edu

Abstract. Single-molecule fluorescent imaging may provide an avenue to efficiently detect the Ba^{++} daughter atom in the decay $^{136}Xe \rightarrow Ba + 2e^-$, and, unambiguously associate the birth point in space within the electron trajectories of the decay event. Chelation of doubly-charged alkaline earth elements such as calcium and barium by certain precursor molecules converts the resulting complex from a non-fluorescent to a fluorescent state. Repeated photo-excitation of a single fluorescent complex reveals both presence and location with high precision. This technique, widespread now in biochemistry, biophysics and biology, may permit a similar discriminating response in a large high-pressure xenon gas TPC for the Ba^{++} ion from xenon double-beta decay. The TPC measures the event time and energy of the two nascent electrons, as well as topology and position in 3-D from their trajectories in the gas. Measurement of the 2-D location of the molecular ion after arrival at the cathode plane permits an association of ion with the event. Demonstration of an efficient, highly specific detection of the barium daughter would provide a long-sought pathway to a background-free result in the search for this decay mode, of central importance for determining the nature of the neutrino.

1. Introduction

Observation of neutrino-less double beta-decay in any of several candidate nuclei will establish that the neutrino is its own anti-particle, a unique property among spin 1/2 particles, and that lepton number is not conserved [1]. Any observation of this decay mode implies a natural and attractive connection to very high masses and a plausible mechanism for the baryon-antibaryon asymmetry of the universe, physics well beyond the Standard Model. The known neutrino mass parameters establish for the inverted mass ordering hierarchy an effective neutrino mass $\langle m_{\beta\beta} \rangle$ range of 20 - 50 meV, corresponding to possible decay lifetimes T in excess of $\sim 10^{27}$ years. Recent results [2] for ^{136}Xe have established lower limits of $\sim 2 \times 10^{25}$ years, corresponding to sensitivities of approximately $150 < \langle m_{\beta\beta} \rangle < 280$ meV. These results are sufficient to rule out a controversial earlier result for ^{76}Ge, but include contaminating backgrounds and make no discovery claims. Such long lifetimes present daunting experimental challenges. To probe deeply into the range of the inverted mass ordering hierarchy, active masses approaching or exceeding ton-scale are necessary.

In addition, substantial technical advances to reduce backgrounds are essential. In a background-free experiment, sensitivity to the neutrino effective mass increases slowly, as $(\text{run-time})^{1/2}$. However, if the probability of even *one* background event is significant, sensitivity evolves toward $(\text{run-time})^{1/4}$. As soon as background is present in an experiment, further running adds marginal sensitivity; further scientific reach is effectively blocked. The natural sensitivity goal is to span the inverted mass ordering range, *i.e.*, to reach ~20 meV. If background is demonstrably zero, only about 1 ton-year is needed. But if backgrounds in the true event class reach 10 events/ton-year, then an exposure of about 100 ton-years is needed for 90% confidence, illustrating the catastrophic impact of background. Quantitatively, one might arguably take *background-free* to mean that the probability P for any background event satisfying all true event criteria in one year of running must be not greater than, say, 6%, or possibly as low as 1%; I'll choose 3%. At the ton scale, P = 3% for background contamination of the energy ROI corresponds to $P \leq 3 \times 10^{-5}$ counts/kg-year. With 1% FWHM energy resolution this corresponds approximately to $P < 1 \times 10^{-6}$ counts/keV-kg-year. No contemporary or planned experiment is able to claim or approach this level of discrimination.

As backgrounds may arise from both external radioactivity and from the allowed two-neutrino double beta-decay, various strategies are needed for mitigation. Heroic radio-purity regimens for materials are essential but have not yet been sufficient. Rejection of background events from two-neutrino double beta-decay will require energy resolution generally not worse than 1% FWHM. Germanium ionization detectors and tellurium oxide bolometric crystals have energy resolutions ten times better this, but have not yet been able to reject adequately all ambient radioactivity.[1] A new method to relieve the daunting future challenges of radio-purity will be most welcome.

The experimental quest, now more than half of a century old since the first insights of Racah, Goeppert-Mayer and Furry in the 1930s, thus remains an exciting but increasingly exacting scientific arena [3]. As a definitive observation of this decay mode would constitute a major discovery, the quality of evidence for a discovery claim must accordingly be very high. A new era has arrived–apparently without much notice–in which future experiments will be very costly in time-scale, careers, and resources and cannot be justified without very robust *a priori* background estimates supporting a credible, nearly background-free capability. The bazaar of technically interesting but scientifically marginal technical approaches explored earlier must yield to a small number of demonstrably robust 'discovery class' experiments.

2. Detection of the daughter atom

Detection of the daughter atom has been long recognized as a possible avenue to realize background rejection, since conventional processes do not introduce a new atom with Z ± 2. For barium, xenon's double-beta decay daughter, attention has been given to spectroscopic features of the singly ionized state, Ba^+, as this atomic configuration permits a sequence of repetitive excitation/de-excitation cycles with red and blue light involving a long-lived triplet D state [4]. Detection of an extended sequence of photons of two colors from a single atom is generally accepted as a robust determination of the presence of Ba^+.

Ba^+ two-color excitation and scintillation is, however, only possible in near-vacuum conditions. D-state quenching prevents measuring both spectroscopic features of single Ba^+ in high pressure or liquid xenon *in situ*. Conversely, the extraction of a single ion from a large mass of liquid or high-pressure gaseous xenon with rapid transport into a low-pressure trap where spectral interrogation becomes effective also appears to be extremely difficult, perhaps intractable, at least to some observers.

In the decay $^{136}Xe \rightarrow Ba + 2e^-$, with or without neutrinos, it is very likely that the daughter barium atom will be highly ionized by the disruptive departure of the nascent electrons from the nucleus. A highly ionized barium atom will strip electrons from nearby neutral xenon atoms until it is unable to de-ionize further. The process halts when barium reaches the doubly ionized state Ba^{++}. At this point,

[1] This is perhaps due to the need to embed a large number of small detectors with relatively high surface-to-volume ratio within an inter-penetrating structural harness, unlike the case of the monolithic xenon detectors.

the second ionization potential of barium, 10.04 eV, is less than the first ionization potential of xenon, 12.14 eV, [5]. Further extraction of electrons from neutral xenon atoms is energetically forbidden. So in pure xenon, the natural state of the equilibrium barium ion is expected to be, instead of singly ionized, doubly charged Ba^{++}. This may turn out to be a great advantage.

3. Single-molecule fluorescent imaging

But there may be another way for sensing the creation of barium, one that may be compatible with high-pressure xenon gas. Single-molecule fluorescent imaging (SMFI) is among the most powerful techniques in contemporary biological sciences. SMFI, developed by physicists, has been adopted and perfected by biologists and chemists.[2] The basic SMFI idea is to interrogate repeatedly a small, optically thin region with blue or near-UV photons that can excite a molecule of interest. Repeated interrogations can provide statistically precise localization down to the few nm^2 level, surpassing the Abbe diffraction limit. Detection of a fluorescent response is made possible by a chromatic shift or a delay in response time relative to a pulsed excitation. Fluorescent response quantum yields approach unity in many cases. Two-photon excitation by IR offers a way to obtain, by filtering, near-perfect rejection of the excitation wavelength. Many texts and references are available [6,7].

A wide variety of chemical and biological molecules has been found or constructed for SMFI. Of particular interest, for two reasons, are calcium-chelating fluorophores. First, doubly-charged calcium ions are fundamental in biology and hence prominent in biochemistry research. Second, since barium and calcium are congeners, what works for Ca^{++} may be relevant for barium. For example, Fluo-3, Fluo-4, and PET-1 are commonly used for calcium chelation. Calcium chelation by Fluo-3 is shown schematically in figure 1. Typical excitation of Fluo-3 is at 488 nm, with response peaking at 550 nm. Quoted response ratios between unchelated and chelated states vary from 60 to more than100. In a dry environment, the response ratio is unknown and may be quite different.

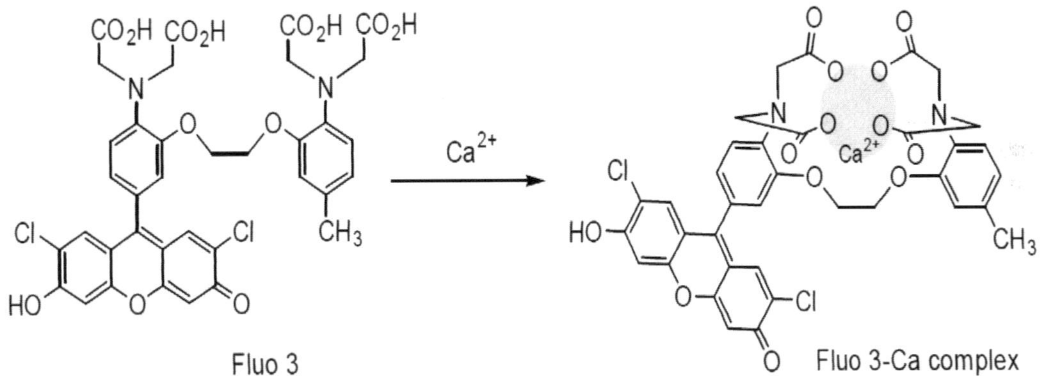

Figure 1. The skeletal formula for Fluo-3 is shown before and after complex formation. The conformal change that Fluo-3 undergoes in chelation with Ca^{++} creates a fluorescent complex. In that state, the fluorescent response increases by a factor of 60 - >100 in the cellular environment. *Image source*: Dojindo Molecular Industries, Inc.

Especially noteworthy is pyrene-stabilized monoazacryptand **1** [8]. This fluorophore is responsive with extremely high specificity to Ba^{++}, ignoring almost completely other alkaline earth doubly charged ions, even Ca^{++}, as well as all alkaline metal ions. Excitation at 342 nm leads to fluorescence in a 330-420 nm band. At this very early stage, with so many unanswered questions, the existence of numerous possible molecular avenues is comforting.

[2] The 2014 Nobel Prize in Chemistry was awarded to three scientists, trained as physicists.

Journal of Physics: Conference Series **650** (2015) 012002 doi:10.1088/1742-6596/650/1/012002

In the presence of the TPC electric field, the barium ion will drift slowly to the cathode plane. The cathode is presents a thin dielectric layer to avoid neutralizing the arriving ion. The search problem is thus reduced from 3-D to a 2-D problem. If the cathode dielectric surface is coated with the incipient fluorophore, and if it is further assumed that the chelation process works in the high-pressure xenon gas, then, exactly one molecular complex becomes responsive to excitation whereas prior to that occurrence there is none. The unsolved challenge is to ascertain by interrogation that this is the case.

The interrogation/response rate in SMFI can exceed 10^5 per second. With the assumptions of Gaussian statistics, negligible photon background, and that calibrations establish an average single ion response to be, say, 100 detected photons, one might naïvely expect that the detection of 100 fluorescence photons from a single site establishes the presence of a single ion with variance $\sigma = 10$. The probability that two or more ions (or zero!) are responding to yield 100 $\pm 3\sigma$ detected photons is, in this idealized case, extremely small. An isolated complex can be interrogated in less than 100 μs. Only those event candidates with energy near the parent-daughter Q-value need detailed scrutiny. However, for an event to be considered as a double beta-decay candidate, it will be necessary to ascertain that exactly one new detected barium ion is now present within the general locus of the event candidate as projected on the cathode plane. Diffusion of the ion during drift, on the order of 1 mm rms, requires a relatively large search area within the projected event locus.

An immediate problem in this scenario is that the un-complexed molecules have to be unresponsive to a very high level if the search area is 'large'. A monolayer presents approximately 10^{10} molecules/ mm^2. A way out of this dilemma, with the addition of some magical thinking, is to imagine that the dielectric surface is a moving belt that transports the ion to a region where only charged molecules are selected. This could be achieved by having the moving belt surface pass by a reversed field region in a convenient location. At this point, all charged complexes would transfer to a co-moving roller where the scan occurs, leaving all the un-complexed ones on the belt. Thus the precursors do not contribute in the actual scan. This scenario is reminiscent of the familiar xerographic copy process, for which nano-xerography has been explored with some success. Some aspects of this are illustrated in figure 2.

The basic idea presented here is an augmentation of the high-pressure xenon gas (HPXe) electroluminescent (EL) TPC concept [9,10]. The HPXe EL TPC provides the desired energy resolution $\delta E/E < 1\%$ FWHM for the two nascent electrons through electroluminescence (EL), a nearly noiseless linear gain method [11]. Two-electron event topology (or single electron, as from γ-rays) is revealed in excellent detail in xenon gas at densities corresponding to ~10 bars. The increasing multiple scattering as the two electrons range out reveals the two-electron topology signature.

4. Perspective

The scale of interest is to achieve sensitivity that spans the inverted mass ordering. This implies a exposure goal of at least 1 ton-year with demonstrably negligible backgrounds. In the author's view, following a path that leads in a ton-scale experiment to a background-dominated result is failure.

The possibility to exploit SMFI is both appealing and admittedly bizarre at first read. A large number of questions arise for this concept, and the exploratory path forward will likely present numerous cul-de-sacs along the way. Nevertheless, the quest to find a technical approach that provides adequate and robust background rejection is compelling. A radically novel approach, if plausibly credible, is more rational to pursue and develop than following familiar paths to certain failure. Perhaps the biochemistry of $^{136}Xe \rightarrow {}^{136}Ba^{++}$ can show us the path to discovery.

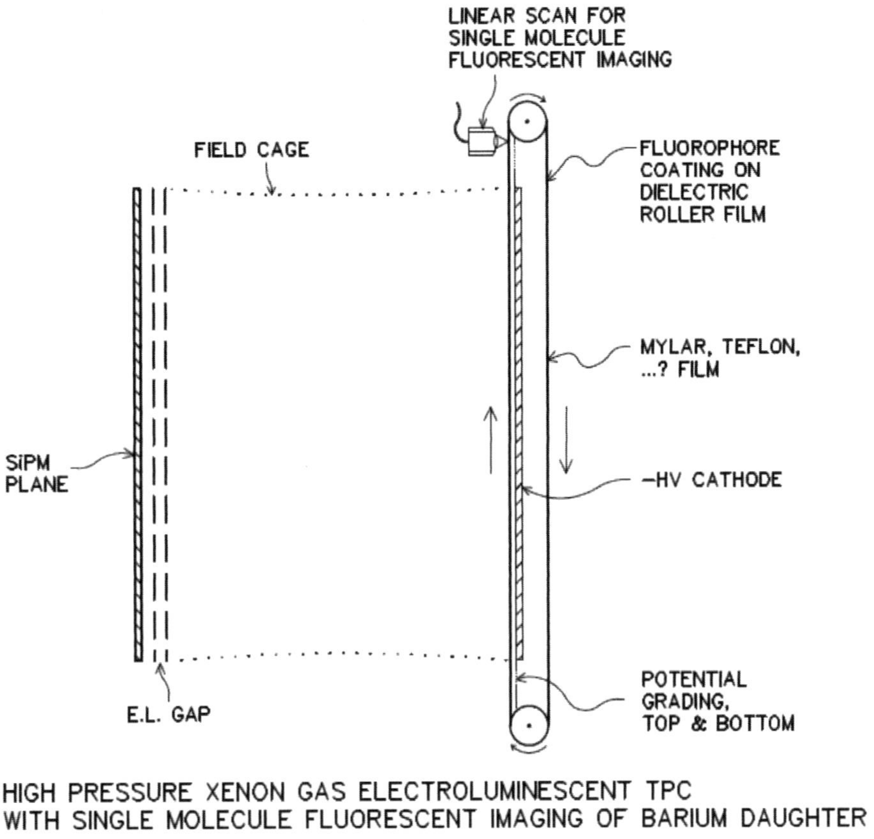

Figure 2. The 3-D search problem for a single barium ion is reduced to 2-D by collection of the ion on a dielectric cathode surface. The cathode surface is a slowly moving belt that transports the complex to a 1-D scanner that senses the latent image with resolution of ~0.1 mm. The primary decay electrons, imaged in 3-D at the anode plane, predict the locus on the belt where, for true events, a barium ion complex should appear.

References

[1] Valle J W F, arXiv:hep-ph/060810v1

[2] Auger M *et al.* (EXO collaboration) arXiv:1205.5608v2 [hep-ex]; Asakura K *et al.* (KamLAND-ZEN collaboration) arXiv:1409.0077v1

[3] Elliott S R and Vogel P, "Double Beta Decay" *Annual Reviews of Nuclear and Particle Science 2002* vol 52 p115-151

[4] Moe M Physical Review C vol 44 (1991) p931

[5] IAU Symposium 210 Uppsala; see IAU Symposium 210

[6] Valeur B and Berberan-Santos M N, *Molecular Fluorescence* 2nd ed Wiley-VCH 2013

[7] Lakowicz J R *Principles of Fluorescence Spectroscopy* 3rd ed Springer 2010

[8] Nakahara Y, Kida T, Naktsuji Y, and Akashi M *A novel fluorescent indicatior for* Ba^{++} *in aqueous micellar solutions* The Royal Society of Chemistry Communications 2004 p224

[9] Nygren D, Nuclear Instruments and Methods in Physics Research vol A 603 (2009) p337-348

[10] Gomez-Cadenas J J (NEXT collaboration) arXiv:1411.2433v1

[11] Alvarez V *et al.* arXiv:1211.4474v1 [physics.ins-det]

Journal of Physics: Conference Series **650** (2015) 012003 doi:10.1088/1742-6596/650/1/012003

Micro-physics simulations of columnar recombination along nuclear recoil tracks in high-pressure Xe gas for directional dark matter searches

Y Nakajima, A Goldschmidt, M Long, D Nygren, C Oliveira, J Renner

Lawrence Berkeley National Laboratory, Berkeley, CA 94720, USA

E-mail: YNakajima@lbl.gov

Abstract. Directional sensitivity is one of the most important aspects of WIMP dark matter searches. Yet, making the direction of nuclear recoil visible with large target masses is a challenge. To achieve this, we are exploring a new method of detecting directions of short nuclear recoil tracks in high-pressure Xe gas, down to a few micron long, by utilizing columnar recombination. Columnar recombination changes the scintillation and ionization yields depending on the angle between a track and the electric field direction. In order to realize this, efficient cooling of electrons is essential. Trimethylamine(TMA) is one of the candidate additives to gaseous Xe in order to enhance the effect, not only by efficiently cooling the electrons, but also by increasing the amount of columnar recombination by Penning transfer. We performed a detailed simulation of ionization electrons transport created by nuclear recoils in a Xe + TMA gas mixture, and evaluated the size of the columnar recombination signal. The results show that the directionality signal can be obtained for a track longer than a few μm in some ideal cases. Although more studies with realistic assumptions are still needed in order to assess feasibility of this technique, this potentially opens a new possibility for dark matter searches.

1. Introduction

Discovering the nature of dark matter is one of the most important goals in Particle Physics. If the dark matter consists of Weakly Interacting Massive Particles (WIMPs), they would occasionally interact with matter on Earth and make low-energy nuclear recoils at O(10) keV. The signature of such nuclear recoils would be a small energy deposition with a preferred direction due to the Solar System's rotation in our galaxy. Most of the experiments with leading sensitivity use liquid noble gas detectors for their scalability and effectiveness of background rejection. In such detectors, however, searches solely rely on the detection of a small energy deposition, and detecting the directional signal is extremely difficult since the track length of a nuclear recoil in liquid is expected to be only O(10) - O(100) nm.

On the other hand, there are many ongoing efforts aimed at WIMP dark matter searches with directional sensitivity using low-pressure ($\sim$100 torr) gaseous TPCs. In such detectors, the track length would be O(mm) and can be detected with imaging devices. A drawback of this kind of detectors is that the density of target gas is typically 10^{-4} times smaller when compared with liquid noble gas detectors, which makes it very hard to achieve competitive sensitivity.

To fill the gap, we are exploring a new method of detecting the direction of nuclear recoils using columnar recombination [1], which would be capable of identifying the direction of a track several μm in length in a high-pressure ($\sim$10 bar) gaseous TPC. The density of target material can be about 100 times larger than that of the low-pressure gaseous TPCs designed for directional dark matter searches. If this technique is realized, it would be the first detector that is capable of detecting the directional signal of nuclear recoil with a target mass large enough to compete with liquid-based TPCs. In order to test the feasibility of this idea, we performed a detailed simulation of transport and recombination of ionization electrons in a high-pressure gaseous medium.

2. Directional sensitivity with columnar recombination

We are exploring directional dark matter searches with high-pressure gaseous Xenon (Xe) TPCs, and the following study is based on this kind of TPC. Xenon is an attractive target medium, because of its high mass and therefore greater predicted spin-independent WIMP scattering cross section, and also for the possibility of searching for neutrinoless double beta decay of ^{136}Xe with the same detector. On the other hand, a similar principle of directional dark matter searches should be applicable to TPCs based on other noble gases, such as Argon or Neon.

2.1. Columnar recombination

If a WIMP particle collides a Xe atom in the detector medium, the recoiled nucleus will excite and ionize other Xe atoms along its track. Figure 1 shows a simplified schematic of this process. Typical observable signals are VUV scintillation light from excited Xe and charge from ionization electrons. The recombination of Xe^+ and e^- produces scintillation light as well, and therefore changes the ratio of scintillation light and charge yields.

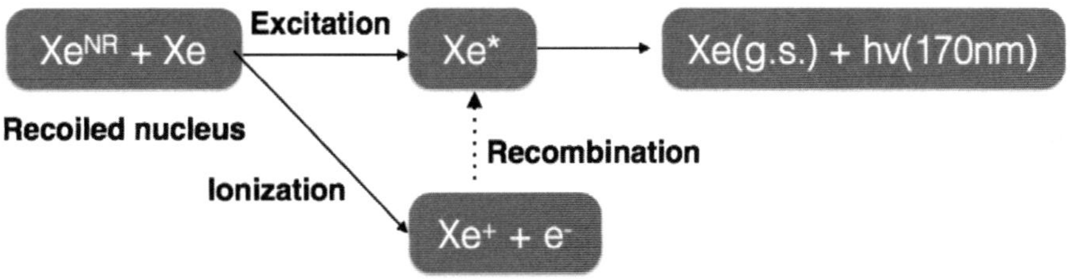

Figure 1. Simplified schematic of interaction of recoiled Xe nucleus in Xe medium

Columnar recombination is the process by which the ionization electrons recombine with other non-parent Xe ions along the track. Its probability depends on the relative angle between the track and the external electric field. If the track is parallel(perpendicular) to the electric field, ionization electrons have more(less) chance to meet Xe ions and therefore recombination would be increased(decreased), as illustrated in Fig. 2. This is an observed phenomena for α-particles [2] in a Xe + TMA gas mixture. The goal of this study is to see if the phenomena can be observed for much shorter tracks of recoiled O(10) keV Xe nucleus.

In order to obtain the directional sensitivity with columnar recombination, the distributions of both ions and electrons from the ionization process need to preserve the directional information while they overlap each other. We describe these requirements in detail below.

2.2. Requirement for ions

Figure 3 shows a simulated distribution of ionized Xe made by 30 keV recoiled Xe nuclei in 10 bar Xe gas, calculated using the Stopping and Range of Ions in Matter (SRIM) simulation

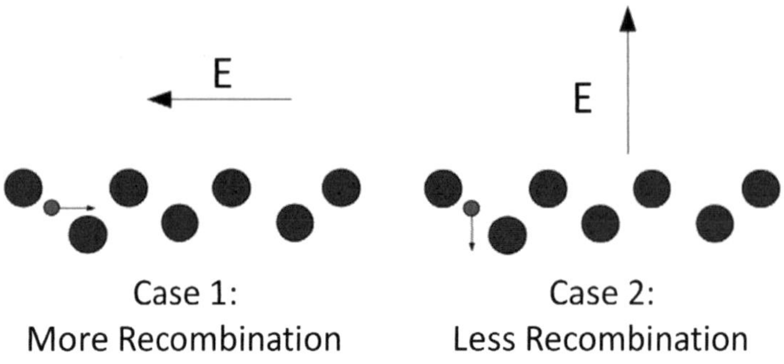

Figure 2. Schematic of columnar recombination. The large brown circles represent ions and the small blue circles represent electrons.

package [3]. As shown in this figure, distribution along the original recoil direction is larger than its transverse direction. The energy-weighted range of 30 keV Xe ions in 10 bar high-pressure Xe gas is expected to be about 2 μm.

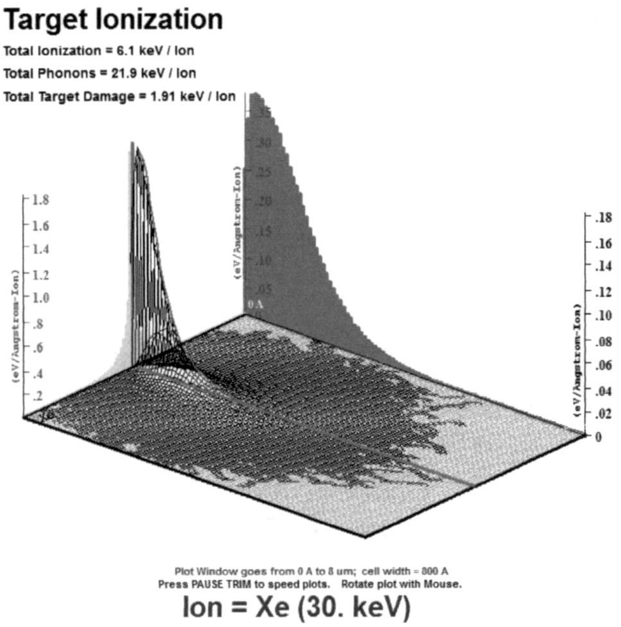

Figure 3. Simulated distribution of ionization made by 10000 Xe ions of 30 keV, injected along the X-axis (the red line), shown on a 8×8 μm plane. The red(green) histogram shows the distribution of ionization projected onto the axis parallel(perpendicular) to the original ion direction. The Xenon gas density is set to $\rho = 0.05$ g/cm^3, which corresponds to about 10 bar at room temperature. This figure was taken from Ref. [1].

Ionization electrons are attracted to ions by their Coulomb potential before they recombine. Therefore, the shape of the Coulomb potential that ions create is one of the factors that determine the directional sensitivity. To quantify this, we evaluate the Onsager radius, r_0, which is the distance at which the Coulomb potential of the electron-ion system is equal to the thermal energy $k_B T$, where k_B is the Boltzmann constant, and T is the temperature. It can be simply calculated as $r_0 = e^2/(\epsilon k_B T)$, where e is the charge of electron, ϵ is the dielectric constant. This r_0 roughly corresponds to the resolution of detecting a distribution of ions by ionization electrons.

Table 1 shows the typical track length of nuclear recoils and the Onsager radius for liquid and gaseous Xe. In the case of liquid Xe, the expected track length would be roughly equal to the Onsager radius, and therefore no significant directionality signal would be detected. On the other hand, in gaseous Xe, the track length would be much longer than the Onsager radius, and it should be visible as a track for ionization electrons. Therefore, it is essential to use gaseous phase Xe in order to utilize the columnar recombination. The density of a gaseous phase detector is lower compared to a liquid phase detector, but the expected detector size that would be necessary to compete with currently leading experiments is still within a realistic scale. For example, the physical size of a 1-ton target mass with a 10 bar high-pressure gaseous Xe detector would be about 20 m^3.

Table 1. Comparison of typical track length due to Xe nuclear recoils of 30 keV kinetic energy and the Onsager radius for liquid and gaseous Xe phases.

	Density	Track length (30 keV NR)	Onsager radius (r_0)
Liquid Xe	3.1 g/cm^3	$\sim$ 50 nm	$\sim$ 50 nm
Gas Xe (10 bar, 300K)	0.05 g/cm^3	$\sim$ 2 μm	$\sim$ 70 nm

2.3. Requirement for electrons

The requirements for electrons are more stringent, as the directional information is further washed out due to the diffusion process. In addition, electrons need to be thermalized in order to efficiently recombine with ions.

Pure Xe would not satisfy this requirement, because it lacks inelastic scattering processes below its first excitation energy of $\sim$7 eV, and it will take an extremely long time to thermalize electrons and cause large diffusion. Therefore, molecular additives with large inelastic cross sections at low energy are needed. We chose Trimethlamine (TMA) as a candidate additive, because it has a large inelastic cross section due to many vibrational and rotational modes, which would efficiently cool down electrons. In addition, it is expected to TMA enhances the ionization signal through the Penning effect ($Xe^*+TMA \rightarrow Xe+TMA^+ + e^-$), which contributes to increased columnar recombination. We will examine more details about the effect of molecular additives in the following section.

3. Micro-physics simulation of recombination

3.1. Simulation setup

In order to understand the electron diffusion process with such molecular additives and to make a realistic estimation of the recombination signal, we performed a micro-physics simulation of electron transport in the gas mixture of Xe and TMA under a uniform external electric field. The simulation is based on Garfield++ [4] and Magboltz [5] with a custom implementation of the recombination process which takes into account the influence of the Coulomb field created by ions and electrons themselves. The electron interaction cross sections with Xe and TMA in Magboltz version 9.01 are used for this work. Only electrons produced via primary ionization are tracked in the simulation, and secondary ionization is not considered. The distribution of initial electron energy, dN/dE, is set to $dN/dE \propto 1/(E^2 + (7.6\ \text{eV})^2)$ with a cut-off at 7 eV, based on Ref. [6]. The temperature was set to room temperature (293 K). An electron was considered to have recombined with an ion if it passed within one de Broglie wavelength of it. We simulated up to 100 electron and ion pairs simultaneously. Simulating more than 100 electron and ion pairs was possible but not practical because of its computational intensiveness.

Journal of Physics: Conference Series **650** (2015) 012003 doi:10.1088/1742-6596/650/1/012003

3.2. Size of electron diffusion

In order to extract directionality information from columnar recombination, the diffusion of ionization electrons must be small enough that their spread is smaller than the track length while they overlap with the ions. Assuming a typical drift velocity of $\sim$O(1) μm/nsec and a track length of a few μm, the duration of the overlap is expected to be a few nsec after the initial ionization.

Figure 4 shows the size of the standard deviation of simulated electron location perpendicular to the electric field, σ_R, as a function of the time since the initial ionization at 10 bar. The results show that the size of diffusion can be significantly reduced with the addition of TMA. With the gas mixture of 10% TMA and 90% Xe, the size of σ_R can be reduced to be $\sim$2 μm for 5 nsec after the initial ionization.

As shown in the right panel of Fig. 4, the diffusion process can be separated into two distinct phases; the initial $\sim$0.1 nsec with quick expansion where the effect of TMA is marginal, and the following relatively stable phase. The time to reach the stable phase tends to be shorter with increased TMA concentration.

This phase transition is due the thermalization process of the electrons. Figure 5 shows the average electron kinetic energy as a function of the time since the ionization. TMA is expected to cool electrons quite efficiently, and the average kinetic energy of electrons would reach the thermal energy of $\sim$0.025 eV within $\sim$0.1 nsec. Therefore, adding TMA to the gas mixture would shorten the initial expansion phase before thermalization and would help to keep ionization electrons close to the ions.

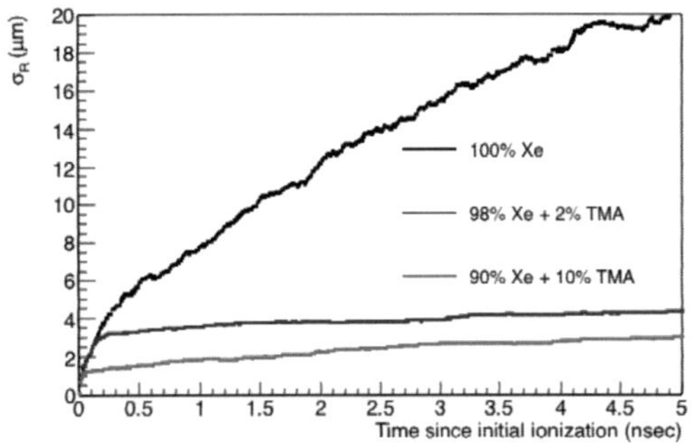

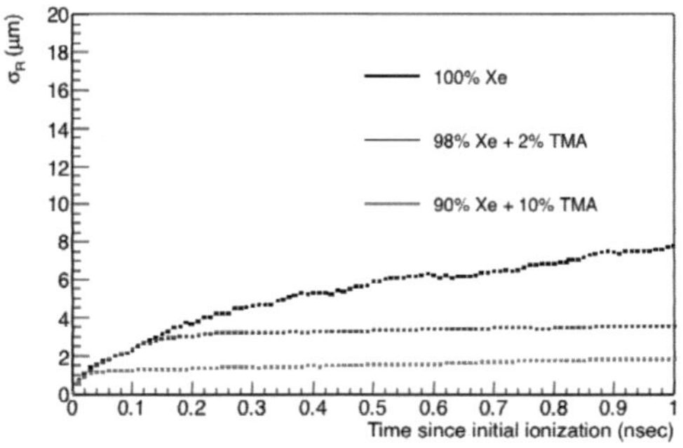

Figure 4. The simulated size of σ_R as a function of time since initial ionization for various Xe + TMA mixtures at a total pressure of 10 bar. Results from pure Xe (black), a 98% Xe and 2% TMA mixture (blue) and a 90% Xe and 10% TMA mixture (red) are shown. The left and right panels show the same data over different horizontal axis ranges.

3.3. Expected recombination signal

The result in the previous section shows that the electrons are expected to be kept within $\sim$4(2) μm of the track for a 2(10)% mixture of TMA at 10 bar total pressure. This suggests that directional sensitivity from columnar recombination can be obtained for tracks longer than $\sim$4(2) μm with a TMA concentration of 2(10)%. The actual recombination probability would be a complex function of both TMA fraction and the strength of the external electric field. To estimate this function, we simulated the transport of 100 electron-ion pairs under the following initial conditions:

- A 1 μm track with 100 equally separated electron-ion pairs aligned either parallel or perpendicular to the external electric field. (10 nm spacing)

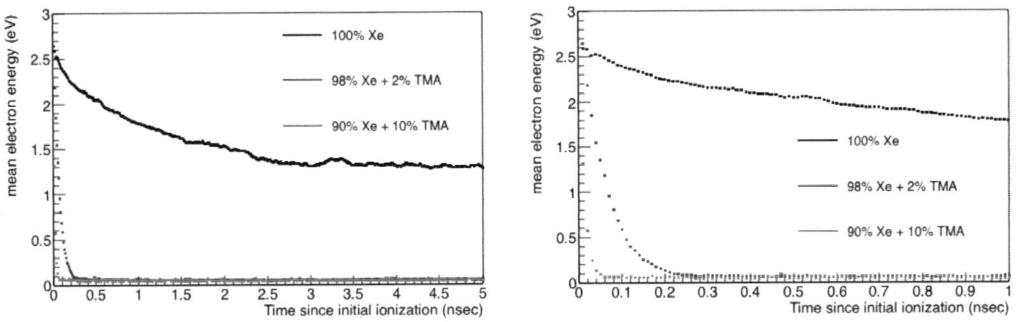

Figure 5. The simulated size of the average kinetic energy of ionization electrons as a function of time since initial ionization for various Xe + TMA mixtures at the total pressure of 10 bar. Results from pure Xe (black), a 98% Xe and 2% TMA mixture (blue) and a 90% Xe and 10% TMA mixture (red) are shown. The left and right panels show the same data over different horizontal axis ranges.

- A 4 μm track with 100 equally separated electron-ion pairs aligned either parallel or perpendicular to the external electric field. (40 nm spacing)

For the first set of simulations with a 1 μm track length, the 10 nm spacing was chosen since it is approximately the same as that of α particles and is a realistic ionization density for a nuclear recoil, even though the track length is shorter than the σ_R and no directionality signal is expected. On the other hand, while the ionization density is less realistic, we expect some directionality signal in the simulations of 4 μm tracks since the track length would be equal to or longer than σ_R for the simulated Xe + TMA mixtures.

Figure 6 shows the simulated recombination fraction from the 1 μm and 4 μm tracks in various Xe + TMA mixtures and electric fields. The simulation predicts significant enhancement of recombination with the addition of TMA for both 1 μm and 4 μm tracks. This recombination fraction naturally decreases as the external electric field gets stronger. As expected, we see no sign of directional sensitivity for the 1 μm tracks in any combination of gas mixtures and external fields. On the other hand, we found that the simulation predicts a statistically significant difference of the recombination fraction between 0-degree (parallel to the external field) and 90-degree (perpendicular to the external field) cases for the 4 μm tracks in the 10% TMA and 90% Xe gas mixture at $E/p > 100$ (V/cm/bar). We also found some directional sensitivity for the 2% TMA and 98% Xe gas mixture at $50 < E/p < 150$ (V/cm/bar). Those are consistent with the expectation based on the size of σ_R described in the previous section.

Although more realistic assumptions and benchmarks with the experimental data are needed, these simulation results show that it may be possible to detect directionality of short tracks of a few μm by utilizing columnar recombination.

4. Discussions

Although the results from the previous section are encouraging, many other things must still be considered to realistically estimate sensitivity to directional information.

First, this study assumed that the initial ionization was aligned on a straight line. However, the actual distribution is expected to be smeared as shown in Fig. 3, which will reduce the directional information.

Another idealization made was the assumption of 100% detection efficiency for both scintillation light and ionization electrons. While detecting ionization electrons at high efficiency is relatively easy, the detection efficiency of primary scintillation light tends to be small and it is $\leq$O(10)% in actual experimental apparatus, because of a typically limited solid angle coverage and the quantum efficiency of photon detectors. Therefore, the uncertainty of the results shown in Fig. 6 would be larger in any realistic detector configuration.

The energy distribution of ionization electrons is also poorly understood. There are several previous measurements on ionization electron energy spectra from Xe [6, 7, 8], but there are no data points below a few eV, which is the most important energy region for columnar recombination. In addition, Penning transfer, if exists, adds a lower energy component to the electron spectrum. This would make a significant impact on the size of the initial expansion of the electron spatial distribution, and hence the size of columnar recombination. Additional inputs from experimental data are extremely important.

Finally, it should also be noted that recent experimental studies revealed that TMA significantly reduces scintillation light yield when it is mixed with Xe gas [9]. Therefore, it is likely difficult to use the gaseous mixture of Xe and TMA for dark matter searches. On the other hand, any molecular additives that have a large inelastic cross section that is similar to TMA would enhance columnar recombination as well. Therefore, it is essential to look for other candidate molecular additives.

Those factors must be taken into account in estimating realistic directional sensitivity. We expect that there will be still many technical challenges in order to realize this idea.

5. Summary

We are exploring a novel idea of detecting the direction of a very short track in high-pressure Xe gas by utilizing columnar recombination. We performed a micro-physics simulation of ionization electrons to study this possibility. The results show that, with sufficient molecular additives such as TMA which efficiently cool electrons, it may be possible to detect the directionality of a few μm track in a high-pressure Xe gas. While more realistic assumptions need to be made in order to evaluate the feasibility of this technique, it is an encouraging start and opens a new possibility for WIMP dark matter searches.

6. Acknowledgments

This work was supported by the Director, Office of Science, Office of Basic Energy Sciences, of the US Department of Energy under contract no. DE-AC02-05CH11231. J. Renner (LBNL) acknowledges the support of a US DOE NNSA Stewardship Science Graduate Fellowship under contract no. DE-FC52-08NA28752.

References

[1] Nygren D 2013 *J.Phys.Conf.Ser.* **460** 012006
[2] Herrera D C *et al.* 2015 *Proceedings of Science* **TIP2014** 057
[3] Ziegler J F The Particle Interactions with Matter URL `http://www.srim.org`
[4] Garfield++ – simulation of tracking detectors URL `http://cern.ch/garfieldpp`
[5] Magboltz – transport of electrons in gas mixtures URL `http://magboltz.web.cern.ch/magboltz/`
[6] Opal C B, Beaty E C and Petterson W K 1972 *Atomic Data* **4** 209–253
[7] Toburen L H 1974 *Physical Review A* **9** 2505–2517
[8] Chaudhry M A 1989 *Physical Review A* **39** 530–536
[9] Nakajima Y *et al.* *In these proceedings*

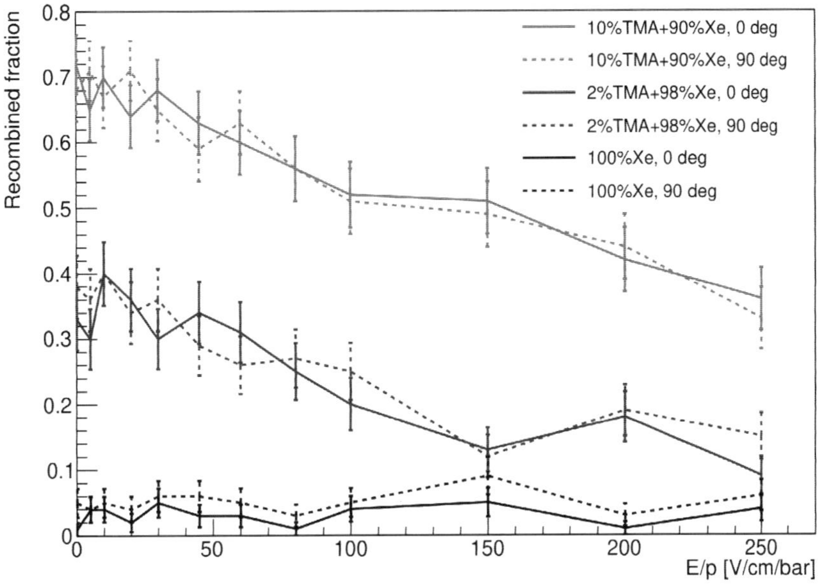

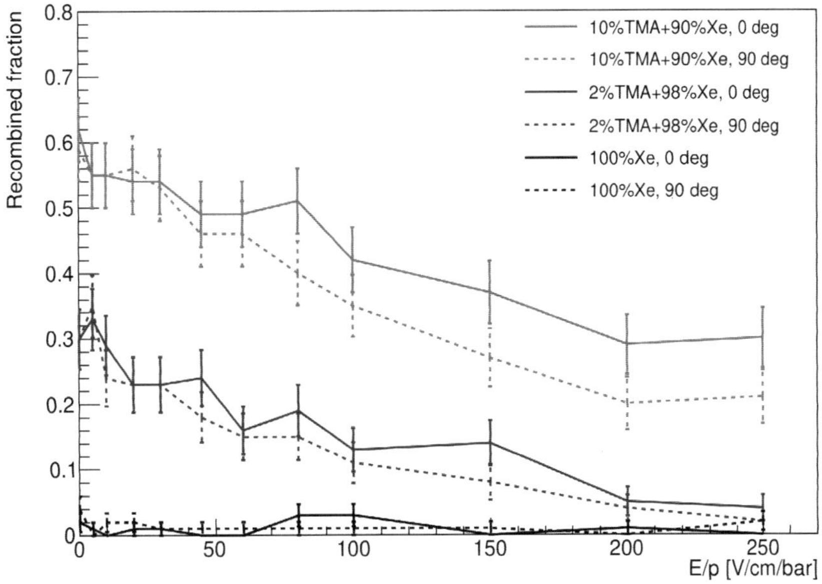

Figure 6. The simulated size of recombination fraction, (total number of recombined electrons) / (total number of initial ionization electrons), from 100 electron-ion pairs in a 1 μm track (top panel) and a 4 μm track (bottom panel). The results for various reduced external electric field and various Xe + TMA mixtures at the total pressure of 10 bar are shown. The solid(dashed) lines shows the results with the initial track parallel(perpendicular) to the external electric fields. The black, blue, and red points are results with pure Xe, a 98% Xe and 2% TMA mixture, and a 90% Xe and 10% TMA mixture, respectively. The error bars show the statistical uncertainty of a single event assuming that both total number of recombination photons and initial ionization electrons are measured with 100% efficiency.

Journal of Physics: Conference Series **650** (2015) 012004 doi:10.1088/1742-6596/650/1/012004

ICARUS T600 - a large Liquid Argon Time Projection Chamber

Jan Kisiel
(on behalf of the ICARUS T600 Collaboration)
Institute of Physics, University of Silesia, Uniwersytecka 4, 40-007 Katowice, Poland
E-mail: Jan.Kisiel@us.edu.pl

Abstract. The Liquid Argon Time Projection Chamber (LAr TPC) is a detection technique very well suited for the investigation of rare signals, including neutrino interactions. The ICARUS T600 is the largest LAr TPC ever built. It has recorded events both, from the CERN to Gran Sasso (CNGS) muon neutrino beam and cosmic rays interactions. Several important results have been achieved thanks to the three-dimensional event imaging and very good detector resolutions both, calorimetric and spatial. Two of them, namely the observation of a free electron lifetime exceeding 15 ms, and the search of LSND related anomalies in the ν_e appearance from the ν_μ CNGS beam will be shortly summarized.

1. The ICARUS T600 detector

Following C. Rubbia's idea of LAr TPC detection technique [1] the ICARUS Collaboration, finalized many years of *R&D* studies with a construction of the T600 detector which, with its 760 t of LAr (476 t active mass) represents the biggest LAr TPC built so far. The collection of both, the ionization electrons at anode wires, and scintillation light by photomultipliers (PMTs) allow to reconstruct, in three dimensions, any ionizing particles with spatial resolution of about 1 mm^3. The ICARUS T600 cryostat is split into two identical modules. Each of them houses two TPCs with a common cathode placed in the middle. The maximum drift length of electrons is equal to 1.5 m. The 500 V/cm uniform electric field drifts electrons with the drift velocity v_d = 1.6 mm/μs towards three anode wire planes (0°, ± 60° with respect to the horizontal direction), 3 mm apart. Thank to appropriate voltages, the signal is recorded in a non-destructive way in the first two *Induction* wire planes, whereas the ionization charge is collected on the last *Collection* plane. Arrays of PMTs, placed behind wire planes, detect prompt 128 nm scintillation light, providing the measurement of the absolute time T_0 of an ionizing event. Both, T_0 and v_d are used to determine the absolute position of the track along the drift coordinate. The ICARUS T600 detector is triggered mainly by the sum of PMTs signals. For CNGS beam events, the trigger is generated when the PMTs sum signal is present in at least one TPC chamber within 60 μs gate opened at the proton extraction time, which is known from the *Early Warning Signal* sent from CERN to LNGS 80 ms before each proton extraction. The coincidence of the PMTs sum signals in two adjacent chambers triggers cosmic ray events. Detailed description of the ICARUS T600 detector can be found in [2] and [3].

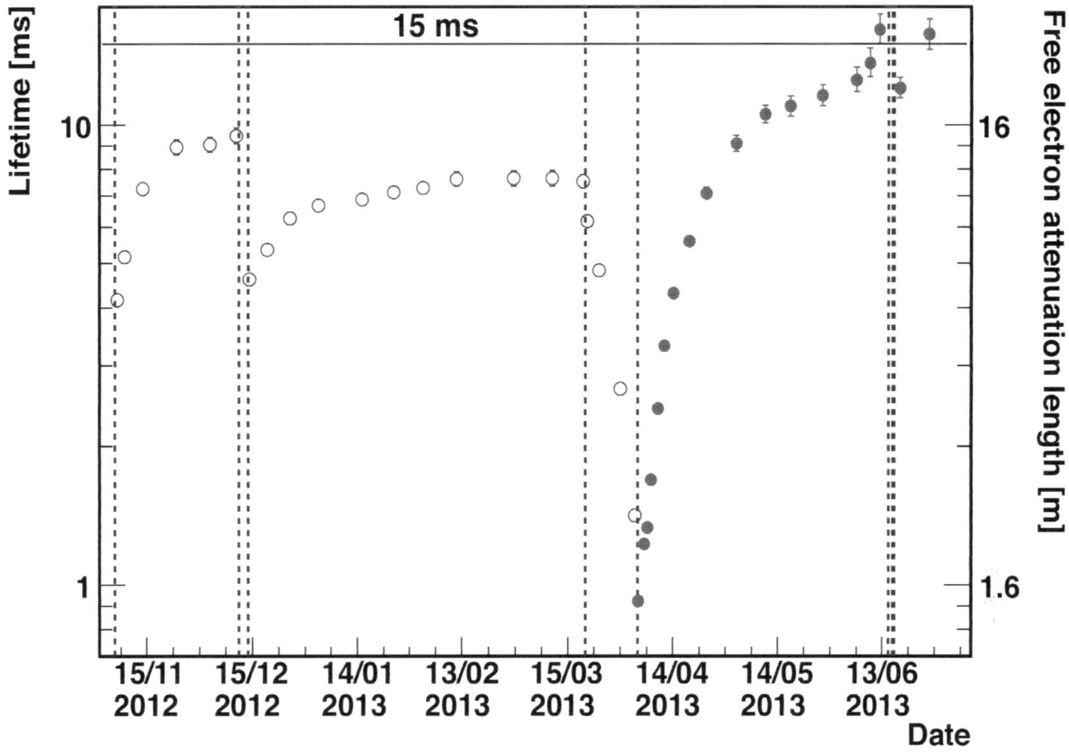

Figure 1. Electron lifetime and free electron attenuation length in the East module of the ICARUS T600 detector for the last period of data taking: full red points corresponds to the measurements with the new pump.

2. Electron lifetime and LAr purity

The operation of large LAr TPC requires to reach and maintain a very high level of LAr purity. Each ICARUS T600 module has been equipped with two gas Argon and one liquid Argon recirculation/purification systems ([3], [4]) which kept the electronegative impurities (mainly O_2, H_2O and CO_2) in LAr at a very low concentration level. The gas Argon, from the cryostat ceiling is continuously drawn, re-condensed, filtered and finally returned to the cryostat, whereas LAr is continuously recirculated by an immersed, cryogenic pumps (6 days for full volume recirculation) and finally re-injected into the cryostat. The attenuation of charge, produced by cosmic-ray muon tracks traversing the detector, was used to determine the electron lifetime. The results for the East cryostat are presented in Fig. 1. In April 2013 the ACD CRYO pump, used in the first two years of data taking, was replaced with a new Barber Nichols BNCP-32C-000 with an external motor. With the new pump, the electron lifetime started rising up, and reached $16.1^{+1.3}_{-1.1}$ ms before the detector stop. This electron lifetime corresponds to a maximum signal attenuation of 6% at the maximum drift distance of 1.5 m in the ICARUS T600 detector. The details of the electron lifetime determination in the ICARUS LAr TPC can be found in [5].

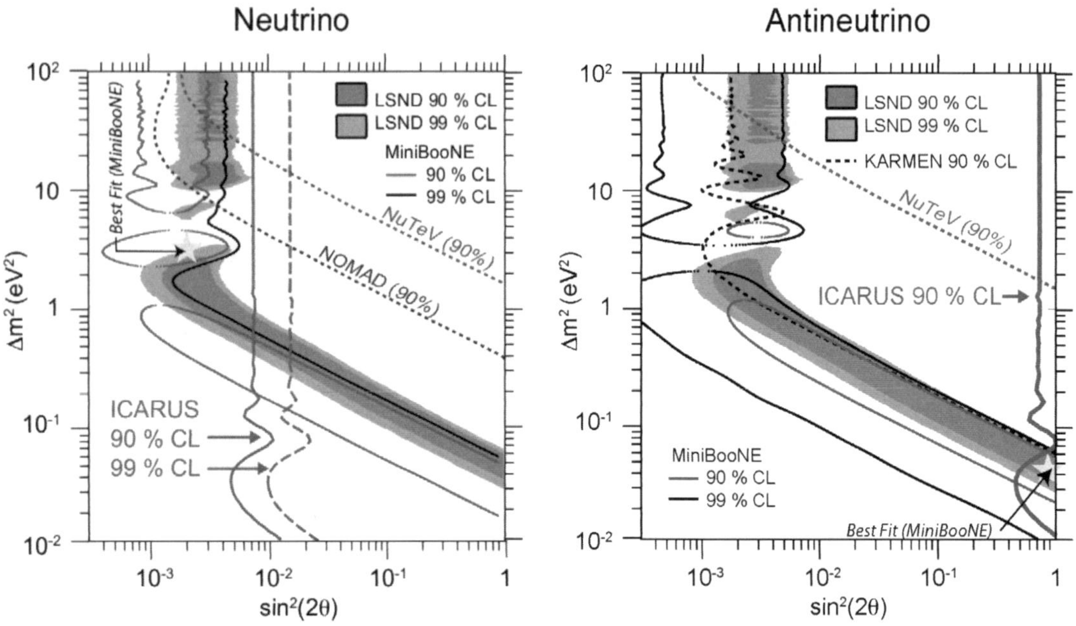

Figure 2. Δm^2 as a function of $\sin^2(2\theta)$ for neutrino (*left*) and antineutrino (*right*) for the main experiments sensitive to the $\bar{\nu}_\mu \to \bar{\nu}_e$ and $\nu_\mu \to \nu_e$ anomalies [6], [7], [8], [9] and [10]. The continuous red line represents ICARUS result, whereas the yellow stars mark the best fit points of MiniBooNE [7].

3. Search for anomalies in the electron neutrino appearance from a muon neutrino beam

The LSND experiment observed an excess of anti-ν_e neutrino events in anti-ν_μ beam: 87.9 $\pm$ 22.4 $\pm$ 6.0 (3.8σ) [6], which was later partly confirmed by MiniBooNE collaboration [7]. These results may imply the existence of a new additional sterile neutrino flavors with the mass-squared difference Δm^2_{new} of about 0.01 to 1.0 eV^2 with a associated value of $\sin^2(2\theta_{new})$, which is highly incompatible with the three-neutrino mixing approach. Thanks to the excellent γ-electron separation in the LAr TPC, the ICARUS Collaboration searched for electron neutrino appearance in the CNGS ν_μ beam in the context of LSND-like anomalies. The first result, based on the study of 1955 neutrino events, corresponding to 6.0×10^{19} POT, was published in [11]. In this proceedings the updated analysis based on 2450 neutrino events, corresponding to the 7.2×10^{19} POT, described in details in [12], is summarized. Visual selection of electron neutrino event candidates in the detector fiducial volume, with an event energy less than 30 GeV (50% reduction of the beam ν_e background and only 15% reduction of the expected signal), was performed. The ν_e event signature required: (1) a single minimum ionizing particle (dE/dx $\leq$ 3.1 MeV/cm, excluding δ-rays) originating from the interaction vertex, over at least 8 wires long, later developing into an electromagnetic shower, and (2) minimum 150 mrad spatial separation from other tracks coming from the vertex at least in one of two transverse views. Taking into account the selection efficiency 0.74 $\pm$ 0.05, the expected number of ν_e events from known sources (residual ν_μ contamination in the ν_μ beam and events due to three neutrino mixing standard model predictions) is equal to 7.9 $\pm$ 1.0 (syst. error only). Six ν_e events have been found. This result is compatible with the expectations due to known sources, giving following limits (weighted for efficiency) on neutrino events due to LSND anomaly: 5.2 (90 % C.L.), or 10.3 (99 % C.L.), with corresponding limits on the oscillation probabilities: P($\nu_\mu \to \nu_e$) $\leq$ 3.85

$\times$ 10^{-3} (90 % C.L.) and $P(\nu_\mu \rightarrow \nu_e) \leq 7.60 \times 10^{-3}$ (90 % C.L.). The exclusion area of the ICARUS experiment is shown in Fig. 2. A 2% muon anti-neutrino contamination is present in the CNGS beam [13]. Assuming that the whole effect is due to $\bar{\nu}_\mu \rightarrow \bar{\nu}_e$, the absence of an anomalous signal gives a limit of 4.2 events at 90% C.L., which corresponds to the limit on the oscillation probability $P(\bar{\nu}_\mu \rightarrow \bar{\nu}_e) \leq 0.32$, shown in Fig. 2 as the small exclusion area.

4. Conclusions

The ICARUS T600 detector had been acquiring data without interruption for more than 3 years, in the underground Gran Sasso laboratory, with both CNGS beam and cosmics, proving the maturity of this detection technique. Remarkable LAr purity had been achieved resulting in the electron lifetime of 16 ms, what is important for next generation experiments. Developed reconstruction algorithms allow to resolve most of the events collected, down to their single components. As an example, analysis of electron neutrino events has been shown. No evidence of oscillation into sterile neutrinos has been found. At the end of 2014 the ICARUS T600 detector has been transported to CERN for overhauling and refurbishment (upgrade of light collection system and electronics, new cold vessel, new passive insulation and new cathode) within the WA104 experiment (details can be found in [14]). The upgraded ICARUS T600, together with MicroBooNE and LAr1-ND detector comprises a combined Fermilab Neutrino program to address the short-baseline anomalies and search for sterile neutrinos.

5. Acknowledgements

I would like to thank the ICARUS Collaboration for the help in preparing my TPC 2014 Conf. talk and this proceedings, as well. This work is supported by the Polish National Science Center: Harmonia (2012/04/M/ST2/00775) funding scheme.

References

[1] C. Rubbia 1977 CERN-EP/77-08
[2] S. Amerio *et al.* (ICARUS Coll.) 2004 *Nucl. Instr. and Meth.* **A527** 329
[3] C. Rubbia *et al.* (ICARUS Coll.) 2011 *Journal of Instrumentation* **6** P07011
[4] C. Vignoli *The ICARUS T600 liquid argon purification system*, Joint 25th ICEC and ICMC Conf., Enschede, The Netherlands, 2014.
[5] M. Antonello *et al.* (ICARUS Coll.) 2014 *Journal of Instrumentation* **9** P12006
[6] A. Aguilar *et al.* (LSND Coll.) 2001 *Phys. Rev.* **D64** 112007
[7] A. Aguilar-Arevalo *et al.* (MiniBooNE Coll.) 2013 *Phys. Rev. Lett.* **110** 161801
[8] B. Armbruster *et al.* (KARMEN Coll.) 2002 *Phys. Rev.* **D65** 112001
[9] P. Astier *et al.* (Nomad Coll.) 2003 *Phys. Lett.* **B570** 19
[10] S. Avakumov *et al.* (NuTeV Coll.) 2002 *Phys. Rev. Lett.* **89** 011804
[11] M. Antonello *et al.* (ICARUS Coll.) 2013 *Eur. Phys. J.* **C73** 2345
[12] M. Antonello *et al.* (ICARUS Coll.) 2013 *Eur. Phys. J.* **C73** 2599
[13] N. Agafonova *et al.* (OPERA Coll.) 2011 *New J. Phys.* **13** 053051
[14] M. Bonesini *these proceedings*

TREX-DM: a low background Micromegas-based TPC for low mass WIMP detection

F.J. Iguaz, J.G. Garza, F. Aznar[1], J.F. Castel, S. Cebrián, T. Dafni, J.A. García, H. Gómez, D. González-Diaz, I.G. Irastorza, A. Lagraba, G. Luzón, A. Peiró and A. Rodríguez

Laboratorio de Física Nuclear y Astropartículas, Universidad de Zaragoza, Spain.

E-mail: `iguaz@unizar.es`

Abstract. Dark Matter experiments are recently focusing their detection techniques in low-mass WIMPs, which requires the use of light elements and low energy threshold. In this context, we present the TREX-DM experiment, a low background Micromegas-based TPC for low-mass WIMP detection. Its main goal is the operation of an active detection mass ~0.300 kg, with an energy threshold below 0.4 keVee and fully built with previously selected radiopure materials. This article describes the actual setup, the first results of the comissioning in Ar+2%iC_4H_{10} at 1.2 bar and the future updates for a possible physics run at the Canfranc Underground Laboratory in 2016. A first background model is also presented, based on Geant4 simulations and a muon/electron discrimination method. In a conservative scenario, TREX-DM could be sensitive to DAMA/LIBRA and other hints of positive WIMPs signals, with some space for improvement with a neutron/electron discrimination method or the use of other light gases.

1. Motivation

The detection of Dark Matter (DM) [1] is one of the open challenges of Astroparticle and Particle Physics for the next years. Evidence for DM is well founded in different observations like the anisotropies in the cosmic microwave background, the distribution of matter in our galaxy or its gravitational effect on visible matter. However, its nature is still unknown but its solution may involve new particles with masses and cross-sections characteristic of the electroweak scale. The most compelling ones are the weakly interactive massive particles (WIMPs), which is a generic class of DM candidates [2], or axions and axion-like particles (ALPs) [3]. From the first type, some positive hints of signals have been reported. The most important one is due to DAMA/LIBRA experiment [4], which has observed an annual modulation (14 cycles) compatible with that expected for galactic halo WIMPs. More recently, CoGeNT [5] and CDMS-II [6] have announced other possible positive signals compatible with a WIMP halo but other analysis [7, 8] have respectively attributed them to a detector effect or an incomplete background model. These hints are in contradiction with the results of other experiments like LUX [9], supossing an spin-independent isosping-invariant WIMP-nucleon coupling and conventional astrophysical assumptions for the WIMP halo.

The current strategy of DM experiments is based on accumulating large target masses of heavy nuclei (like Xenon), keeping low background levels by a systematic radiopurity control of

[1] Present address: Centro Universitario de la Defensa, Universidad de Zaragoza, Zaragoza, Spain

all components and an enhancement of the electron/neutron discrimination methods. However, this strategy is not suitable for low WIMPs masses as for heavy nuclei this translates into very low nuclear recoil energies, and discrimination methods normally fix a relatively high energy threshold (1-10 keV). Given that low WIMP masses are invoked to explain the reported hints, there is now strong interest in exploring new strategies better focused on the low mass range. For instance, CDMSlite experiment [10] has recently reached an energy threshold as low as 90 eVnr in Germanium detectors.

Gaseous detectors may also play an important role in future experiments, as they can reach sub-keV energy threshold ($\sim$ 100 eV) and have access to richer topological information. Most of these experiments (like DRIFT [11] and MIMAC [12]) are focused on directional detection of DM [13], aiming to exploit the relative inhomogeneity of the WIMP incoming momenta, with a maximum expected in the direction of the Cygnus constellation. This stream could be observed by a low-pressure gaseous detector, as it will effectively select nuclear recoils from electrons [14] and it could determine their direction and sense of the nuclear recoil tracks [15]. However, these last features fastly degrade for long drift distances, which may limit the scalability of this type of experiments.

In this context, the TREX-DM experiment proposes another strategy based on high gas pressures, even if neutron/electron discrimination could be less effective, but keeping a low energy threshold. TREX-DM is a low background Micromegas-based TPC for low-mass WIMP detection and will profit all developments made in Micromegas technology [16, 17, 18], as well as in the selection of radiopure materials [19, 20], specially in CAST [21] and NEXT-MM [22] projects. Its main goal is the operation of an active detection mass $\sim$0.300 kg with an energy threshold below 0.4 keVee (as already observed in [21]) or lower. Another experiment with a similar approach is SEDINE [23], a Spherical Proportional Counter filled with a neon-helium mixture at high pressure and that has reported an energy threshold as low as 0.1 keVee.

This article describes the actual setup of TREX-DM and the first results of the comissioning during 2014. In a second part, a first background model of the experiment is presented, as well as its sensitivity to low-mass WIMPs. We conclude with a summary and some prospects.

2. Description and comissioning

The actual setup (see figure 1) is composed of a high purity copper vessel, with an inner diameter of 0.5 m and a length of 0.5 m. The vessel's thickness is 6 cm, so that it could shield most external radiation if combined with an external 10 cm thick lead layer. The vessel contains two active volumes (a in the design), separated by a central copper cathode (b). At each side there is a field cage (d) that makes uniform the drift field along the 19 cm between the cathode and the detector. Each Micromegas detector (e) is screwed to a copper base, which is then attached to the vessel's inner walls by means of four columns. The gas enters the vessel by a feedthrough at the bottom part (h) and comes out by another one at the top part (i). The calibration system consists of a plastic tube entering in the bottom part (h), which allows to calibrate each side at four different points (c) with a ^{109}Cd source, emitting X-rays of 22.1 (K_α) and 24.9 keV (K_β). Finally, the vessel can be pumped from the feedthrough (i) at the top to reduce the outgassing rate from the inner walls.

The design of the two Micromegas detectors is a modified version of the CAST-MM one. Each detector is on a Printed Circuit Board (PCB) of 1.6 mm thickness. The active surface of 25×25 cm^2 is divided in squared pads of 400 μm length and separated by a pitch of 500 μm. Pads are alternatively interconnected in X and Y directions (400 strips per direction) through metallized holes, which are rooted into four connectors prints at the sides of the PCB. Instead of using a leak-tight PCB as in MIMAC detector [24], a flat cable is linked to each connector

Journal of Physics: Conference Series **650** (2015) 012005 doi:10.1088/1742-6596/650/1/012005

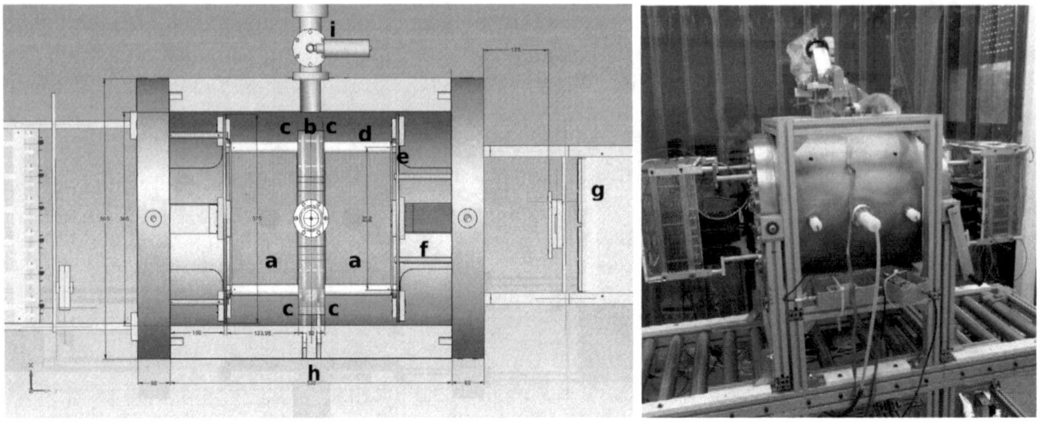

Figure 1. Left: Design of the TREX-DM detector. Its different parts are described in detail in the text: active volumes (a), central cathode (b), calibration points (c), field cage (d), Micromegas detector and support base (e), flat cables (f), AFTER-based electronics (g), gas system (h) and pumping system (i). Right: A view of the experiment during the comissioning.

footprint by means of a SAMTEC connector (GFZ 300 points). The connectivity is assured by four screws, which also join two 1 cm thick lead pieces and two 1 cm-thick copper containers to shield the contamination of SAMTEC connectors [20]. Each flat cable goes out from the vessel by means of a copper feedthrough.

Figure 2. Three views of TREX-DM Micromegas detectors. Left: a bulk detector installed at its support base with its four flat cables already linked to it by four Samtec connectors. Center: Flat cables come out from the vessel by their corresponding feedthroughs. Right: Flat cables are linked to their interface cards, which are then connected to the FECs boards.

The operation principle is the same as in any Micromegas-based TPC. An event interacts in the active volume and releases some electrons, which drift toward the Micromegas plane. Electrons are then amplified in a gap of 128 μm and the charge movement induces signals both at the mesh and the strips. The mesh signal is extracted from the vessel by a low-voltage feedthrough and then decoupled from the high voltage (powered by a CAEN N470A) by a capacitance. It is then consecutively amplified by two CANBERRA preamplifier and amplifier and later recorded by a Tektronix DS5054B oscilloscope. In parallel, strips pulses come out from the vessel by four flat cables. Each of them is linked to an interface card which distributes signals to the entrance connectors of an AFTER (ASIC For TPC Electronics Readout)-based FEC board [25, 26]. Each board has four AFTER ASICs that collect and sample the strip signals continuously at 50 MHz in 512 samples per channel, recording a window of ~10 μs, which is longer than the maximum drift time of charges created in the active volume. The readout electronics is triggered by the negative component of the mesh's amplified bipolar pulse. At that moment, the analog data from all channels is digitized by an ADC converter. Finally, a pure digital electronics card, the FEM board, gathers the ADC data, performs the pedestal substraction and sends it to the DAQ system by means of a standard network. The XZ and YZ views of an event are reconstructed combining the strips pulses, whose temporal position gives the relative z position, and the decoding of both the detector and the interface card. Some examples of reconstructed events are shown in figure 3.

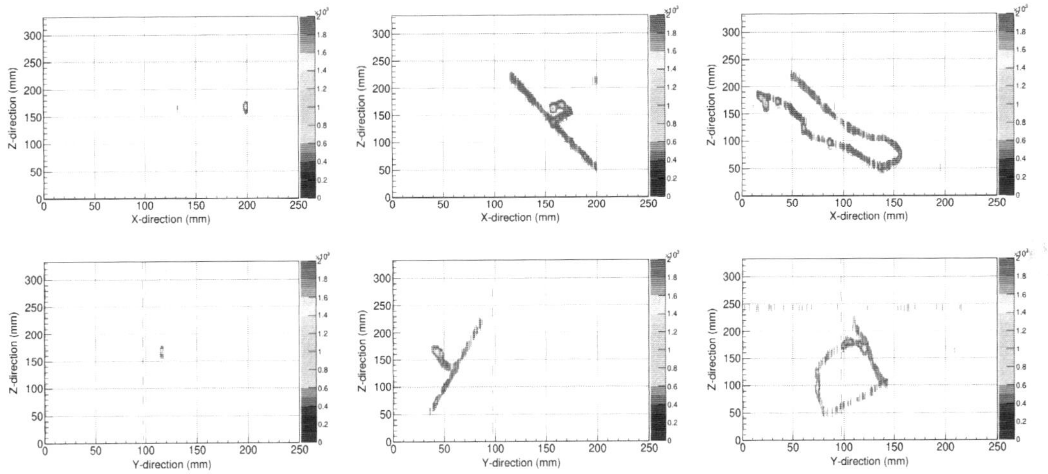

Figure 3. The XZ (top) and YZ views (bottom) of three different events detected during the comissioning of TREX-DM experiment: a low energy X-ray (left), a muon crossing the active volume which generates a δ-ray (center) and an electron with a long twisted track and a final big energy deposition or blob (right). The color scale is proportional to the deposited charge and all x, y and z-axis are in mm.

The TREX-DM prototype is part of the wider scope ERC-funded project called TREX [27], that since 2009 is devoted to R&D on low background TPCs and their potential applications in axion, double beeta decay and dark matter experiments. Work on the TREX-DM prototype started 2012 with the first designs and it is now being commissioned at the TREX lab at Zaragoza. Most of the components have been validated: the leak-tightness of all feedthroughs has been verified as their leak rate is below 10^{-5} mbar l/s; the vessel is leak-tight and can keep pressures up to 10 bar; the drift cage has been tested at high voltage in argon-based mixtures up to 10 bar; the gas flow and pressure, the temperature and the high voltage suppliers are continuously monitored by a slow control programmed in Python and based on Arduino cards [28]; and the first signals were observed at the end of 2014 when the TPC was operated in Ar+2%iC_4H_{10} at 1.2 bar. As described in last section, some components, like the detector or the electronics, will be replaced during 2015 and are only used to validate the experimental design.

The general performance of Micromegas detectors was studied in these conditions. They showed a large plateau at the electron transmission curve, compatible with other bulk detectors.

However, the maximum gain reached before the spark limit (10^3) was far from values reached by bulk detectors (10^4) and the energy resolution at 22.1 keV (~20% FWHM, see figure 4) was modest. The low gain may be explained by the low quantity of quencher in gas [29] and might be improved at higher pressures if Ar+2%iC_4H_{10} is then the optimum mixture [30]. Energy resolution is clearly degraded by noise conditions during the data-taking, due to the presence of a 1 MHz frequency. This noise was present both at mesh and strips and may be produced by a bad grounding at some point of the electronics noise. Nevertheless, the energy threshold was situated at 1 keV.

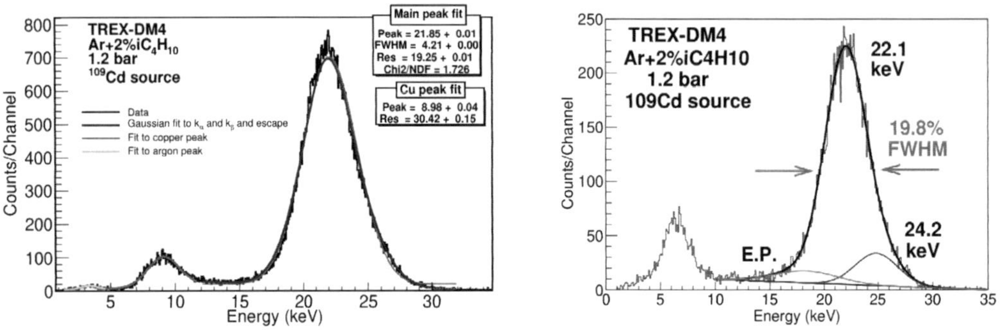

Figure 4. Energy spectrum respectivelly generated by mesh (left) and strips signals (right) when one of the Micromegas detectors was irradiated by a ^{109}Cd source in Ar+2%iC_4H_{10} at 1.2 bar. The energy resolution at 22.1 keV of both spectra has been calculated fitting the spectrum to four gaussians, corresponding to the Kα and K_β lines of the source and their escape peaks (E.P.). The fluorescence lines of iron (at 6.4 keV, emitted from the mesh) and copper (8 keV, from the vessel) are also present in both spectra. The energy threshold is situated at 1 keV.

3. A first background model of TREX-DM

The sensitivity of TREX-DM has been studied creating a first background model of the experiment, if it were installed in the Canfranc Undeground Laboratory (LSC). We have studied two light gas mixtures at 10 bar: Ar+2%iC_4H_{10} and Ne+2%iC_4H_{10}; with an active mass of 0.300 and 0.160 kg respectively. These gases are good candidates to detect WIMPs of masses below 20 GeV. However, one of argon's isotopes (^{39}Ar) is radioactive (β-decay, Q = 565 keV) with a long half-life (239 yr) and may limit the sensitivity of any argon-based experiment. This isotope is not present in argon if it is extracted from undeground sources, as it only appears by cosmogenic activation [31]. We have also simulated the main radioactive isotopes of all the components considering their measured activities [19, 20] and the cosmic muon flux in Canfranc, at a depth of 2450 m.w.e. In the case of the Micromegas detectors and Samtec connectors, we have considered the activities of their radiopure alternative. The external gamma flux has not been considered as its contribution may be supressed by an external shielding.

The simulation of the high pressure TPC can be divided into two blocks. The first one covers all the physical processes involved in the passage of gamma-rays and charged particles through matter, and is based on the Decay0 [32] and the Geant4 simulation packages [33]. In this last program, the actual geometry has been implemented as shown in figure 5 (left). The second block describes the generation of electrons in the gas, the diffusion effects during the drift to the readout plane, the charge amplification in the Micromegas gap and the generation of signals both at mesh and strips. A detailed description can be found in [34]. Some minor changes in data flow have been made to implement a two-volumes geometry, to accelerate the simulation of diffusion process and to implement the AFTER-based electronics. The resulting data has the same format as DAQ data, so as both real and simulated data may be analized by the same routines. As an example, we show in figure 5 (right) the expected energy spectrum when illuminating one side by a ^{109}Cd source situated at a calibration point.

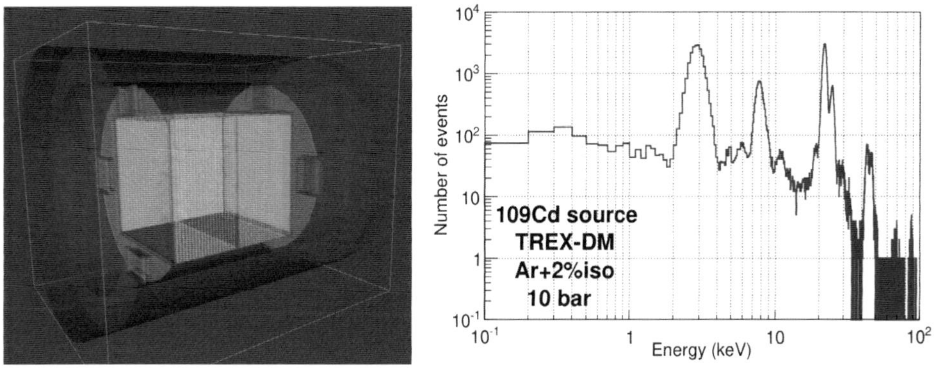

Figure 5. Left: A view of the TREX-DM geometry implemented in Geant4. The cylindrical copper vessel (orange volumes) contains a circular base with four shielded boxes (dark grey surface with four yellow boxes), two active volumes (in light grey), the field cage and degrador (white walls) and a central cathode. Right: Expected energy spectrum in Ar+2%iC_4H_{10} at 10 bar when illuminating one side by a ^{109}Cd source situated at a calibration point.

A modified version of the analysis done in CAST [35] has been used in the background model of TREX-DM. This analysis is optimized in discriminating low energy X-ray from external gammas and cosmic muons and is based on likelihood functions generated by the X-rays' cluster features of a calibration source. Fixing a total 80% signal efficiency, the expected background level for an argon-based gas at 10 bar is 2×10^2 keV^{-1} kg^{-1} day^{-1}, dominated by the ^{39}Ar isotope. If this contribution could be eliminated, the background level will be ~1 keV^{-1} kg^{-1} day^{-1} and will be limited by the copper vessel and connectors, as shown in figure 6 (left). This last value is similar to the one expected in a neon-based mixture. Supossing a 0.4 keVee energy threshold and a conservative background level of 10^2 keV^{-1} kg^{-1} day^{-1}, TREX-DM experiment could be sensitive to a relevant fraction of the low-mass WIMP parameter space (see figure 6, right) including the regions invoked in some interpretations of the hints of the DAMA/LIBRA and other hints of positive WIMPs signals with an exposure of 1 kg-year. In the near-future, we will study the z-dependency of cluster features and we will implement a neutron/electron discrimination, as those made in [14, 36] at low gas pressures.

4. Summary and prospects

The up-to-now strategy of Dark Matter experiments has been focused on accumulating large target masses of heavy nuclei target and reducing the background level by a selection of all components and powerful discrimination methods. However, this approach is not adequate for low-mass WIMPs, as light elements and a low energy threshold are required. In this context, we present the TREX-DM experiment, a low background Micromegas-based TPC for low-mass WIMP detection. Its main goal is the operation of an active detection mass ~ 300 g, with an energy threshold of 0.4 keVee or below and fully built with previously selected radiopure materials.

The experiment consists of a copper vessel divided into two active volumes, each of them equipped with a field cage and a bulk Micromegas 25×25 cm^2 bulk detectors. Signals are

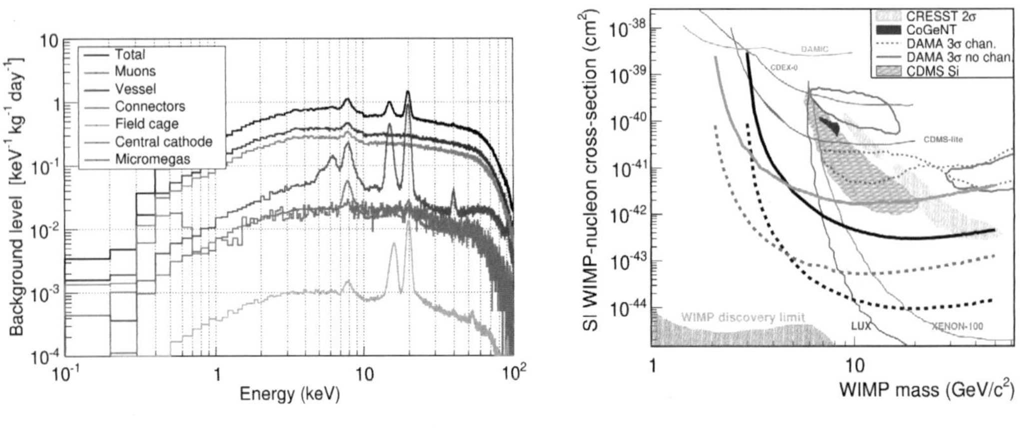

Figure 6. Left: Background spectrum expected in TREX-DM experiment (black line) if operated in Ar+2%iC_4H_{10} at 10 bar in absence of any ^{39}Ar isotope and installed at LSC. The contribution of the different simulated components is also plotted: external muon flux (red line), vessel contamination (blue line), connectors (magenta line), field cage (green line), central cathode (brown line) and Micromegas detector (purple line) have also been included. Right: WIMP parameter space focused on the low-mass range. Filled regions represent the values that may explain the hints of positive signals observed in CoGeNT, CDMS-Si, CRESST and DAMA/LIBRA experiments. The thick lines are preliminary sensitivity of TREX-DM supossing a 0.4 keVee energy threshold and two different hypothesis on background and exposure: 100 (solid) and 1(dashed) keV^{-1} kg^{-1} day^{-1}, and 1 and 10 kg-year respectively, and for both argon- (black) and neon-based mixtures (green).

extracted from the vessel by flat cables and are read by an AFTER-chip based electronics. Each side can be calibrated at four different points by a ^{109}Cd source and the setup is also equipped with other auxiliar systems (gas, pumping, slow-control). The actual setup is being comissioned and the Micromegas detectors have been characterized in Ar+2%iC_4H_{10} at 1.2 bar. Both detectors showed a large plateau at the electron transmission curve, compatible with other bulk Micromegas detectors. However, the maximum gain reached before the spark limit (10^3) and the energy resolution at 22.1 keV (~20% FWHM) are modest. We have attributed these results to a low quantity of quencher in gas and the presence of a high frequency noise. Nevertheless, the energy threshold was situated at 1 keV.

In the near-term, some improvements will be made in the setup like a better detector's grounding or a filter for the cathode voltage to remove its important contribution to noise observed at higher voltages. After these changes, Micromegas detectors will be characterized in Ar+2%iC_4H_{10} up to 10 bar and in other light gases like neon and helium, studying both the energy threshold and the features of low energy X-ray clusters. In parallel, we are developing a large bulk Micromegas detector fully built in radiopure materials. This detector could be also read by AGET electronics [37], which may further reduce the energy threshold to values near 100 eV. These improvements will be commissioned during 2015, so as the detector may be installed at the LSC during 2016 for a possible physics run.

TREX-DM experiment could be sensitive to a relevant fraction of the low-mass WIMP parameter space (see figure 6, right) including the regions invoked in some interpretations of the hints of the DAMA/LIBRA and other hints of positive WIMPs signals, for a 0.4 keVee energy threshold and a background level of $\sim 10^2$ keV^{-1} kg^{-1} day^{-1}, as obtained in a first background model. This result is based on the Geant4 simulation of the radioactivity of the detector components, the signal response of a Micromegas-based TPC and a modified version of CAST analysis, used to discriminate low energy X-rays from muons. A lower background level is expected with a neutron/electron discrimination method, even if it will be less effective at high pressures.

We dedicate this work to the memory of our dear IRFU/SEDI colleagues Marc Anfreville and Michel Boyer, who passed away during 2014.

Acknowledgments

We acknowledge the Micromegas workshop of IRFU/SEDI for bulking our detectors and the Servicio General de Apoyo a la Investigación-SAI of the University of Zaragoza for the fabrication of many mechanical components. We acknowledge support from the European Commission under the European Research Council T-REX Starting Grant ref. ERC-2009-StG-240054 of the IDEAS program of the 7th EU Framework Program and the Spanish Ministry of Economy and Competitiveness (MINECO) under grants FPA2011-24058 and FPA2013-41085-P. F.I. acknowledges the support from the *Juan de la Cierva* program of the MINECO.

References

[1] Baudis L 2012 *Physics of the Dark Universe* **1** 94-108.
[2] Lee B W and Weinberg S 1977 *Phys. Rev. Lett.* **39** 165.
[3] Baker K *et al.* 2013 *Annalen Phys.* **525** A93-A99.
[4] Bernabei R *et al.* 2013 *Euro. Phys. J. C* **73** 2648.
[5] Aalseth C E *et al.* 2011 *Phys. Rev. Lett.* **107** (2011) 141301.
[6] Agnese R *et al. Phys Rev. Lett* **211** (2013) 251301.
[7] Davis J H, McCabe C and Boehm C, *JCAP* **1408** (2014) 014.
[8] Agnese R *et al.* 2014 arXiv:1410.1003.
[9] Akerib D S *et al.* 2014 *Phys. Rev. Lett.* **112** 091303.
[10] Agnese R *et al.* 2014 *Phys. Rev. Lett.* **112** 041302.
[11] Daw E *et al.* 2012 *Astr. Phys.* **35** 397.
[12] Santos D *et al.* 2012 *EAS Publication Series* **53** 25.
[13] Ahlen S *et al.* 2010 *Int. Jour. Mod. Phys. A* **25** 1.
[14] Billard J, Mayet F and Santos D, 2012 *JCAP* **07** 020.
[15] Billard J, Mayet F and Santos D 2012 *JCAP* **04** 006.
[16] Giomataris Y *et al.* 1996 *Nucl. Instrum. Meth. A* **376** 29.
[17] Giomataris I *et al.* 2006 *Nucl. Instrum. Meth. A* **560** 405.
[18] Andriamonje S *et al.* 2010 *JINST* **5** P02001.
[19] Cebrián S *et al.* 2011 *Astropart. Phys.* **34** 354.
[20] Aznar F *et al.* 2013 *JINST* **8** C11012.
[21] Aune S *et al.* 2014 *JINST* **9** P01001.
[22] Alvarez V *et al.* 2014 *JINST* **9** P03010.
[23] Gerbier G *et al.* 2014 arXiv:1401.7902.
[24] Iguaz F J *et al.* 2011 *JINST* **6** P07002.
[25] Baron P *et al.* 2008 *IEEE Trans. Nucl. Sci.* **55** 1744.
[26] Baron P *et al.* 2010 *IEEE Trans. Nucl. Sci.* **57** 406.
[27] http://gifna.unizar.es/trex/.
[28] Peiró A 2013 University of Zaragoza *Preprint:* http://zaguan.unizar.es/record/12149.
[29] Iguaz F J, Ferrer-Ribas E, Giganon A and Giomataris I 2013 *JINST* **7** P04007.
[30] Cebrian S *et al.* 2013 *JINST* **8** P01012.
[31] Acosta-Kane D *et al.* 2008 *Nucl. Instrum. Meth. A* **587** 46.
[32] Pokratenko O A *et al.* 2000 *Phys. At. Nucl.* **63** 1282.
[33] Agostinelli S *et al.* 2003 *Nucl. Instrum. Meth. A* **506** 250.
[34] Tomas A 2013 University of Zaragoza *JINST* **TH** 001.
[35] Garza J G *et al.* 2013 *JINST* **8** C12042.
[36] Iguaz F J *et al.* 2012 *Phys. Proc.* **37** 1079.
[37] Anvar S *et al.* 2011 *IEEE NSS/MIC* 745.

Direct Search for Dark Matter with DarkSide

P Agnes[1], **T Alexander**[2], **A Alton**[3], **K Arisaka**[4], **H O Back**[5], **B Baldin**[6], **K Biery**[6], **G Bonfini**[7], **M Bossa**[8], **A Brigatti**[9], **J Brodsky**[5], **F Budano**[10], **L Cadonati**[2], **F Calaprice**[5], **N Canci**[5], **A Candela**[7], **H Cao**[5], **M Cariello**[11], **P Cavalcante**[7], **A Chavarria**[12], **A Chepurnov**[13], **A G Cocco**[14], **L Crippa**[9], **D D'Angelo**[9], **M D'Incecco**[7], **S Davini**[15], **M De Deo**[7], **A Derbin**[16], **A Devoto**[17], **F Di Eusanio**[5], **G Di Pietro**[9], **E Edkins**[18], **A Empl**[15], **A Fan**[14], **G Fiorillo**[14], **K Fomenko**[19], **G Forster**[2], **D Franco**[1], **F Gabriele**[7], **C Galbiati**[5], **A Goretti**[5], **L Grandi**[12], **M Gromov**[13], **M Y Guan**[20], **Y Guardincerri**[6], **B Hackett**[18], **K Herner**[6], **E V Hungerford**[15], **Al Ianni**[7], **An Ianni**[5], **C Jollet**[21], **K Keeter**[22], **C Kendziora**[6], **S Kidner**[23] **V Kobychev**[24], **G Koh**[5], **D Korablev**[19], **G Korga**[15], **A Kurlej**[2], **P X Li**[20], **B Loer**[5], **P Lombardi**[9], **C Love**[25], **L Ludhova**[9], **S Luitz**[25], **Y Q Ma**[20], **I Machulin**[27,28], **A Mandarano**[10], **S Mari**[10], **J Maricic**[18], **L Marini**[10] **C J Martoff**[25], **A Meregaglia**[21], **E Meroni**[9], **P D Meyers**[5], **R Milincic**[18], **D Montanari**[6], **M Montuschi**[7], **M E Monzani**[26], **P Mosteiro**[5], **B Mount**[22], **V Muratova**[16], **P Musico**[11], **A Nelson**[5], **S Odrowski**[7], **M Okounkova**[5], **M Orsini**[7], **F Ortica**[29], **L Pagani**[11], **M Pallavicini**[11], **E Pantic**[4,30], **L Papp**[23], **S Parmeggiano**[9], **R Parsells**[5], **K Pelczar**[31], **N Pelliccia**[29], **S Perasso**[1], **A Pocar**[2], **S Pordes**[6], **D Pugachev**[27], **H Qian**[5], **K Randle**[2], **G Ranucci**[9], **A Razeto**[7], **B Reinhold**[18], **A Renshaw**[4], **A Romani**[29], **B Rossi**[5,14], **N Rossi**[7], **S D Rountree**[23], **D Sablone**[15], **P Saggese**[7], **R Saldanha**[12], **W Sands**[5], **S Sangiorgio**[3]**2**, **E Segreto**[7], **D Semenov**[16], **E Shields**[5], **M Skorokhvatov**[27,28], **O Smirnov**[19], **A Sotnikov**[19], **C Stanford**[5], **Y Suvorov**[4], **R Tartaglia**[7], **J Tatarowicz**[25], **G Testera**[11], **A Tonazzo**[1], **E Unzhakov**[16], **R B Vogelaar**[23], **M Wada**[5], **S Walker**[14], **H Wang**[4], **Y Wang**[20], **A Watson**[25], **S Westerdale**[5], **M Wojcik**[31], **A Wright**[5], **X Xiang**[5], **J Xu**[5], **C G Yang**[20], **J Yoo**[6], **S Zavatarelli**[11], **A Zec**[2], **C Zhu**[5], **and G Zuzel**[31]

[1] APC, Université Paris Diderot, Sorbonne Paris Cité, Paris 75205, France
[2] Amherst Center for Fundamental Interactions and Physics Department, University of Massachusetts, Amherst, MA 01003, USA
[3] Physics and Astronomy Department, Augustana College, Sioux Falls, SD 57197, USA
[4] Physics and Astronomy Department, University of California, Los Angeles, CA 90095, USA
[5] Department of Physics, Princeton University, Princeton, NJ 08544, USA
[6] Fermi National Accelerator Laboratory, Batavia, IL 60510, USA
[7] Laboratori Nazionali del Gran Sasso, Assergi (AQ) 67010, Italy
[8] Gran Sasso Science Institute, L'Aquila 67100, Italy
[9] Physics Department, Università degli Studi and INFN, Milano 20133, Italy
[10] Physics Department, Università degli Studi Roma Tre and INFN, Roma 00146, Italy
[11] Physics Department, Università degli Studi and INFN, Genova 16146, Italy

[12] Kavli Institute, Enrico Fermi Institute and Dept. of Physics, University of Chicago, Chicago, IL 60637, USA
[13] Skobeltsyn Institute of Nuclear Physics, Lomonosov Moscow State University, Moscow 119991, Russia
[14] Physics Department, Università degli Studi Federico II and INFN, Napoli 80126, Italy
[15] Department of Physics, University of Houston, Houston, TX 77204, USA
[16] St. Petersburg Nuclear Physics Institute, Gatchina 188350, Russia
[17] Physics Department, Università degli Studi and INFN, Cagliari 09042, Italy
[18] Department of Physics and Astronomy, University of Hawai'i, Honolulu, HI 96822, USA
[19] Joint Institute for Nuclear Research, Dubna 141980, Russia
[20] Institute of High Energy Physics, Beijing 100049, China
[21] IPHC, Université de Strasbourg, CNRS/IN2P3, Strasbourg 67037, France
[22] School of Natural Sciences, Black Hills State University, Spearfish, SD 57799, USA
[23] Physics Department, Virginia Tech, Blacksburg, VA 24061, USA
[24] Institute for Nuclear Research, National Academy of Sciences of Ukraine, Kiev 03680, Ukraine
[25] Physics Department, Temple University, Philadelphia, PA 19122, USA
[26] SLAC National Accelerator Laboratory, Menlo Park, CA 94025, USA
[27] National Research Centre Kurchatov Institute, Moscow 123182, Russia
[28] National Research Nuclear University MEPhI (Moscow Engineering Physics Institute), 115409 Moscow, Russia
[29] Chemistry, Biology and Biotechnology Department, Università degli Studi and INFN, Perugia 06123, Italy
[30] Physics Department, University of California, Davis, CA 95616, USA
[31] Smoluchowski Institute of Physics, Jagiellonian University, Krakow 30059, Poland
[32] Lawrence Livermore National Laboratory, 7000 East Avenue, Livermore, CA 94550

E-mail: `pagnes@in2p3.fr`

Abstract. The DarkSide experiment is designed for the direct detection of Dark Matter with a double phase liquid Argon TPC operating underground at *Laboratori Nazionali del Gran Sasso*. The TPC is placed inside a 30 tons liquid organic scintillator sphere, acting as a neutron veto, which is in turn installed inside a 1 kt water Cherenkov detector. The current detector is running since November 2013 with a 50 kg atmospheric Argon fill and we report here the first null results of a Dark Matter search for a (1422 $\pm$ 67) kg.d exposure. This result correspond to a 90% CL upper limit on the WIMP-nucleon cross section of 6.1 $\times 10^{-44}$ cm^2 (for a WIMP mass of 100 GeV/c^2) and it's currently the most sensitive limit obtained with an Argon target.

1. Introduction

Our knowledge of the energy balance of the Universe is derived only by indirect observations. We know that the baryonic matter (the so called luminous matter) only accounts for roughly the 5% of the energy content, while Dark Energy and Dark Matter are estimated to provide the larger contributions, which account for the 68% and the 27% respectively (according to the recent results of the Planck experiment).

The first hypotheses on the existence of Dark Matter date back to the beginning of the 20th century and they are presently well supported by several indirect observations. In spite of this, the knowledge of the nature of these particles, not predicted by the Standard Model, is extremely poor. One of the most favored candidate is known as WIMP, an acronym for Weakly Interactive Massive Particle. These particles are supposed to have masses in the GeV-TeV range, to not interact strongly nor electromagnetically, but only through gravitational and weak forces. The current upper limit on the WIMP-nucleon cross section, extremely low, is 7.6×10^{-46} cm^2 for 33 GeV WIMP mass at 90% CL [1].

The indirect search for WIMPs can be performed by looking for ordinary decay products of WIMP-WIMP annihilation in the Universe or by producing WIMPs in collider experiments;

the direct detection channel can be exploited by searching for nuclear recoils induced by elastic scattering of WIMPs on ordinary nuclei.

A detector for the direct detection must satisfy some requirements: a large active mass; low energy thresholds, since a typical nuclear recoil only deposits less than 100 keV; strong background suppression. Such a detector can be designed to collect the ionization charges, the scintillation light or the phonons produced inside the sensitive medium by the recoiling nucleus. Several experiments are currently running around the world, exploiting different targets and detection techniques, and many among them are designed to access to more than one of those quantities at the same time.

2. Direct Detection of Dark Matter in Noble Liquids

Noble liquids are suitable for direct detection of Dark Matter: they are dense, inexpensive and easy to be purified (a detector can be scaled up to large volumes) and they have high ionization and scintillation yields (roughly one electrons every 20 eV and 40k photons/MeV respectively).

When a charged particle interacts within noble liquids, it looses energy by both ionization and excitation, according to the stopping power of the interacting particle. The excited atoms de-excite producing a prompt light signal in the UV range (scintillation light, called S1 in the following). Electrons and ions produced by ionization can be separated by means of an electric field. A fraction of free electrons, however, undergo to the recombination process with ions produced along the particle track. The recombined atom is in an excited state and the de-excitation increases the S1 signal. The recombination effect is larger at higher ionization densities, and hence stronger for nuclear recoils with respect to electronic recoils. As a result, nuclear and electronic recoils can be discriminated with a rejection factor of the order of 10^2-10^3 [2, 3].

Liquid noble gases experiments, thanks to the double phase TPC technique, are suitable for measuring both the scintillation and the ionization components: the ionization electrons are drifted up to a gaseous layer, lying on the top of the liquid noble volume, and extracted by means of an applied electric field. During the extraction, a second light emission occurs (called S2), thanks to the electro-luminescence effect, proportional to the number of ionization electrons.

The main two noble liquid targets used in currently running experiments are Argon and Xenon. The predicted interaction cross section is slightly different, being larger at low WIMP masses for Xenon. Xenon is denser and highly radio-purer with respect to Argon. Further, the Xenon technology is more advanced and already provided the most stringent limits on the WIMP-nucleon cross section.

The delay of the Argon technology with respect to the Xenon based one is due to the non negligible content of ^{39}Ar in atmospheric Argon. ^{39}Ar is a β emitter, with a Q-value of 565 keV and half life of about 269 years. The typical activity of atmospheric Argon, due to cosmic rays activation, is of the order of 1 Bq/kg and this always prevented the built of a large detector for rare events experiments. The problem can be solved thanks to the recent development of the underground Argon (UAr) extraction technique. This Argon is depleted in ^{39}Ar and a depletion faction larger than 150 has been achieved, as shown in Figure 1. A mass production of underground Argon seems also to be possible at affordable price for future large scale detectors.

Liquid Argon has the advantage of an extremely high discrimination power thanks to the Pulse Shape Discrimination (PSD). The excited Argon atom has two states, with two different decay time constant (the single state, with $\tau_1 \sim 7$ ns and the triplet state with $\tau_2 \sim 1600$ ns). While a nuclear recoil mostly excites the fast state, a typical electronic recoil mostly excites the slow one. Thus, simply looking at the fraction of S1 light that occurs in the first tens of nanoseconds of the signal itself, it is possible to discriminate between nuclear and electronic recoils up to a factor 10^8 [5]. In Liquid Xenon the two time constants are similar ($\tau_1 \sim 22$ ns and $\tau_2 \sim 45$ ns) and the implementation of this technique is prevented [6].

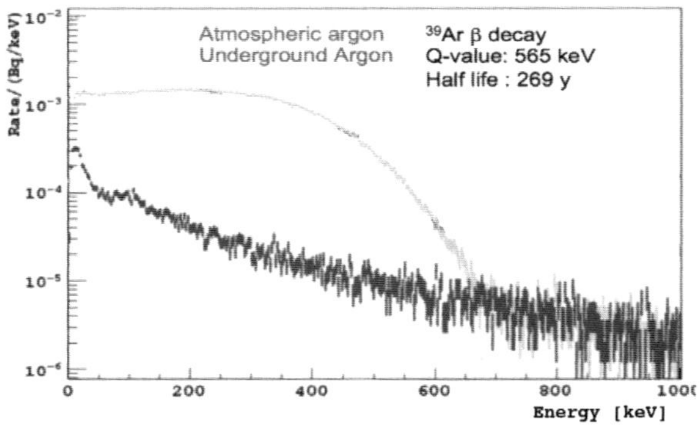

Figure 1. Comparison between the atmospheric Argon (green) and the underground Argon (blue) spectra, as measured at the KURF underground laboratory. The depletion factor results to be > 150 [4].

3. The DarkSide experiment

3.1. The DarkSide program

The DarkSide goal is a background free experiment with a multi-ton scale double phase liquid Argon TPC. In order to accomplish such an ambitious result, the DarkSide collaboration is proceeding through a staged approach. The first prototype (Darkside-10), built in Princeton and running until 2013, proved the stability of the detector and a light yield of about 9 photo-electrons/keV was measured. In the current configuration (Darkside-50) the active mass of the detector has been increased from 10 kg to 50 kg. The full experimental setup includes also a spherical liquid scintillator detector acting as neutron veto and surrounding the TPC, and a water Cherenkov detector for vetoing muons, hosting the sphere.

In spite of the current atmospheric Argon fill and the calibration purposes, the DarkSide-50 first physics results are remarkable and will be discussed in the following.

It is currently under discussion the design of the future phase detector, with a more competitive physics potential. The veto sphere presently installed is already designed in order to host a larger cryostat, so the active mass can easily be scaled up to few tons. The commissioning of the future detector is foreseen to start after the end of the current phase (expected in 2017).

3.2. The DarkSide detector

The detector is installed underground, at *Laboratori Nazionali del Gran Sasso* (LNGS, in Italy), under 3400 m.w.e of rock working as a shield from cosmic rays and it has been realized with highly radiopure materials.

The core of the DarkSide experiment is the double phase liquid Argon time projection chamber, 36 cm diameter × 36 cm height, and filled with roughly 46.7 kg of Liquid Argon (LAr). Two arrays of 19 photo-multipliers are pointing to the center of the volume from the top and from the bottom surfaces (two quartz windows). On the top of the liquid, a 1 cm height gas region is created by heating the LAr. A uniform electric field (200 V/cm) is maintained along the vertical axis of the cylinder and a stronger electric field is present is the gas region (2800 V/cm) for the extraction of ionization electrons.

All the internal surfaces of the TPC are reflective and coated with TPB (ThetraPhenylButadiene), a wavelength-shifter required in order to convert the 128 nm LAr scintillation light in visible one, to match the photocathode sensitivity.

The cryostat is placed inside a 4 m diameter sphere, filled with an organic Liquid Scintillator and equipped with 110 PMTs (8 inches), acting as a neutron veto. The solution is made by 50% Pseudocumene (PC) and TriMetylButadiene (TMB), the latter being a molecule, loaded with Boron, with a very high neutron capture cross section. Also PPO in 5 g/l concentration is added

Journal of Physics: Conference Series **650** (2015) 012006 doi:10.1088/1742-6596/650/1/012006

to reduce the light quenching. The mean lifetime of neutrons inside this solution is of the order of 2 μs and the veto efficiency has been estimated to be higher than 99.9%. The main goal of the neutron veto is the rejection of WIMP-like interactions (nuclear recoils) produced inside the TPC by radiogenic and cosmogenic neutrons; moreover is designed to work actively, not only shielding the TPC from the environment, but also measuring the real neutron background.

The Neutron Veto sphere is then placed inside a 10 kton water tank, with 80 PMTs (8 inches) installed on the side and on the bottom, acting as a Cherenkov detector for the surviving cosmic muons at the depth of the Laboratories. A sketch of the three nested detectors is shown in Figure 2.

Figure 2. The nested detector system of DarkSide-50. The outermost gray cylinder is the Water Cherenkov Detector, the sphere is the Liquid Scintillator Veto, and the gray cylinder at the center of the sphere is the LAr TPC cryostat.

4. The First Results

4.1. Detector calibration and stability

The calibration of the detector has been realized with the insertion of ^{83}Kr inside the Argon circulation loop, a common practice for noble liquids experiment. This radio-nuclide emits two low energy gammas (for a total deposit of 41.5 keV) and has a mean life of 1.8 hours. The position of the 41.5 keV peak over the ^{39}Ar β-spectrum allows to measure the light yield of the detector: 7.9 ± 0.4 photo-electrons/keV without the electric field and ~ 7.0 photo-electrons/keV at 200 V/cm.

The stability of the detector response can also be evaluated selecting the events populating the 41.5 keV peak. While the maximum electron drift time, for the 200 V/cm electric filed, is set to 375 μs ($v_{\text{drift}} \sim 0.93$ mm/μs), the measured electron lifetime is larger than 5 ms. The internal non-uniformity both in terms of light and electrons collection have been evaluated as well.

For the neutron expectation band, it is safe to adopt the results of the SCENE experiment [8], a calibration experiment designed to study the nuclear recoils in LAr with a neutron beam and a small TPC. The reduced dimensions of the TPC are convenient in order to obtain a clean sample of single scattering events and nuclear recoils have been studied for different neutron energies and electric fields.

A calibration campaign with neutron (AmBe) and gamma (^{57}Co, ^{133}Ba and ^{137}Cs) sources at different energies is currently ongoing.

Journal of Physics: Conference Series **650** (2015) 012006 doi:10.1088/1742-6596/650/1/012006

4.2. Background rejection

A WIMP interacting inside the sensitive volume is expected to hit a nucleus and to produce a nuclear recoil. As already mentioned, the main tool for rejecting electron recoils that trigger the TPC is the the Pulse Shape Discrimination (up to a factor 10^8). Exploiting the S2/S1 ratio will increase the rejection power by an additional factor $10^2 \div 10^3$.

The most dangerous source of WIMP-like background is represented by cosmogenic and radiogenic neutrons. Some of these events, those with multiple interaction inside the TPC, can be rejected since they produce a multiple ionization signal. The neutrons interacting only once in the TPC are likely to be captured inside the 4π liquid scintillator surrounding veto. A capture on ^{10}B results in the production of ^{7}Li and α particle. The α energy is 1.47 MeV, quenched to $\sim$ 50 keV. With a branching ratio of 94%, a 480 keV γ is also emitted. The measured light yield in the veto scintillator ($\sim$ 0.52 pe/keV) is large enough to detect the α also when no γ is emitted.

The number of cosmogenic neutrons that penetrate the veto undetected is negligible in a multi-year DarkSide-50 exposure (from calculation). Concerning the internal radioactivity, the major source of neutrons are the PMTs (to be replaced in a future detector by cleaner ones) and the total expected yield is about 100 n/y. From Geant4 based simulation, only 5×10^{-4} of them are expected to interact once in the TPC and to escape the veto without leaving any detectable signal (< 30 pe). This fraction can vary by 20%, because of the large uncertainty on the quenching factor of the α's.

Finally, a fiducialization is applied to the active volume, in order to prevent contamination from α surface emissions (from raw materials qualification, they are expected to be $< 10/\mathrm{m}^2$day), removing events that are originated within 2 cm from the walls. After this cut, the fiducial volume is reduced to 36.9 kg of LAr.

4.3. Extrapolation of the limit

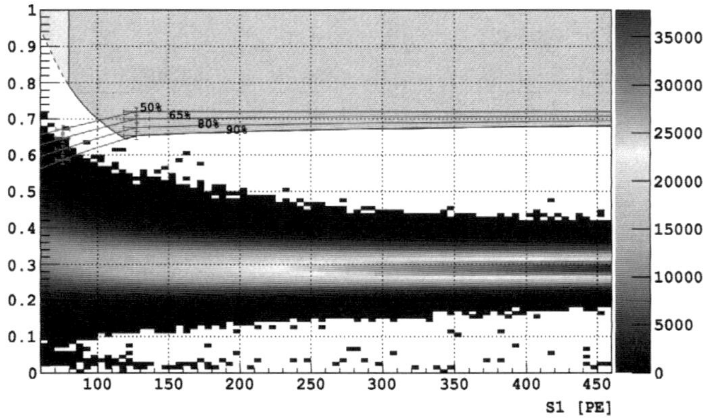

Figure 3. Distribution of the events in the scatter plot of S1 vs. f90 after all quality and physics cuts. Shaded blue with solid blue outline: Dark Matter search box in the f90 vs. S1 plane. Percentages label the f90 acceptance contours for nuclear recoils drawn connecting points (shown with error bars) determined from the corresponding SCENE measurements.

The first result, corresponding to a total lifetime of 47.2 days, is shown in Figure 3 [7]. The plot vertical axis corresponds to the PSD parameter (f90 or the fraction of S1 signal that occurs in the first 90 ns of the signal itself). The f90 parameter is known to be $\sim$ 0.3 for electronic recoils and $\sim$ 0.7 for nuclear recoils. The selected energy window for nuclear recoils extends from roughly 40 to 200 keV$_{\mathrm{nr}}$, namely 8 to 40 keV$_{\mathrm{ee}}$ (electron recoil equivalent).

The Dark Matter search box shown in the plot is obtained by intersecting the 90% nuclear recoil acceptance line from the SCENE experiment with the curve corresponding to a leakage of ^{39}Ar events of 0.01 events/(5-pe bin). The leakage curves are obtained by fitting the f90 distributions for any fixed energy according to the Hinkley model [10].

Journal of Physics: Conference Series **650** (2015) 012006 doi:10.1088/1742-6596/650/1/012006

There are no events in the upper part of the plot, but a large number of electronic recoils (mainly due to the ^{39}Ar) populate the lower region. This result has been obtained with atmospheric Argon runs and the collected statistics corresponds to roughly 20 years of DS-50 run with Depleted Argon.

A first limit on the WIMP-nucleon cross section can be derived from these data, even if a physics result was not the main goal of this calibration campaign. The limit, compared in Figure 4 with the current best results from LUX (a LXe detector) and other experiments, correspond to 6.1 $\times 10^{-44}$ cm^2 (90% CL) for 100 GeV WIMP mass and it is currently the most stringent one obtained with a Liquid Argon target.

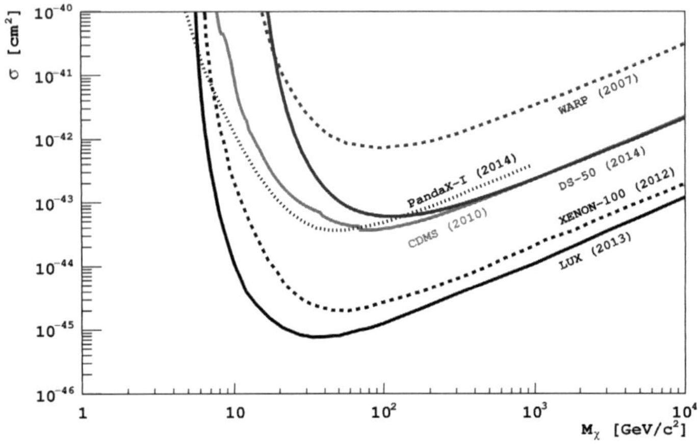

Figure 4. Spin-independent WIMP-nucleon cross section 90% C.L. exclusion plot for the DarkSide-50 atmospheric Argon campaign (Solid Blue Curve) compared with results from LUX [1] (Solid Black Curve) and other experiments.

5. Conclusions and outlook

The DarkSide-50 detector is successfully running and taking data since November 2013. The calibration campaign is almost concluded and the data available so far (corresponding to 50 days lifetime) provided a first physics result.

This result confirms the effectiveness of the Pulse Shape Discrimination: the present exposure of (1422 $\pm$ 67) kg.d with atmospheric Argon, corresponding to 215000 kg.d of running with underground Argon, proves that DarkSide-50 could run for two decades with underground Argon and be free of ^{39}Ar background. In the meanwhile, a first batch of 150 kg of UAr has been shipped to LNGS and the insertion of the liquid inside the cryostat is expected at the end of February 2015.

The design of the next phase detector is currently under discussion. The expected sensitivity for the WIMP-nucleon cross section will be of the order of 10^{-47} cm^2 in a few years exposure, thanks to a ton-scale active mass of underground Argon.

References

[1] D. S. Akerib et al. (LUX Collaboration),Phys. Rev. Lett. 112, 091303 (2014).
[2] Benetti et al. (ICARUS) 1993.
[3] The WARP Collaboration, arXiv:0701286.
[4] arXiv:1204.6011
[5] WARP Astr. Phys 28, 495 (2008)
[6] arXiv.org:1106.2209
[7] The DarkSide Collaboration, arXiv:1410.0653.
[8] T. Alexander et al. (SCENE Collaboration), Phys. Rev. D 88, 092006 (2013).
[9] P. Cushman et al. (APS Community Summer Study 2013), arXiv:1310.8327.
[10] The DEAP Collaboration, arXiv:0801.1531v4.

Journal of Physics: Conference Series **650** (2015) 012007 doi:10.1088/1742-6596/650/1/012007

The LAr1-ND Experiment

Nicola McConkey, for the LAr1-ND collaboration

Department of Physics and Astronomy, University of Sheffield, Hicks Building, Hounsfield Road, Sheffield, S3 7RH, UK

E-mail: n.mcconkey@sheffield.ac.uk

Abstract.

LAr1-ND is a Liquid Argon Time Projection Chamber (LArTPC) based experiment which will operate in the Booster Neutrino Beam (BNB) at Fermilab, at a distance of 110m from the target. The 112 t active volume detector will make precision measurements of the neutrino interaction cross section in argon, as well as forming the near detector for the short-baseline neutrino (SBN) programme at Fermilab. The $4\,\text{m} \times 4\,\text{m} \times 5\,\text{m}$ TPC consists of 2 anode planes, each composed of 3 wire planes, with a central cathode. In combination with the MicroBooNE and ICARUS-T600 detectors, this experiment provides a powerful opportunity to understand observed neutrino anomalies, and has the potential to make precision measurements of oscillations to sterile states.

In addition to the physics goals, the detector technology developed and used in this experiment will form an important step towards the future detectors for long-baseline neutrino experiments. The detector design will be discussed in the context of its physics potential, and contribution to future technologies.

1. Introduction

The Fermilab Short Baseline Neutrino (SBN) program includes the construction of a Liquid Argon Near Detector, LAr1-ND [1] in the Booster Neutrino Beam (BNB), in a new enclosure at 110m from the source. LAr1-ND will form the near detector of a three detector system, with MicroBooNE [2] and ICARUS-T600 [3] as the intermediate and far detectors, as illustrated by figure 1. This system of LArTPCs forms an experiment capable of definitively addressing existing anomalies in neutrino physics and making precision measurements of high-Δm^2 neutrino oscillations through both appearance and disappearance searches.

Leveraging the advanced design work performed for LBNE and the very recent experience of the MicroBooNE detector construction, the LAr1-ND project has the potential to move forward quickly.

Due to the high event rate of neutrino interactions at the near location, significant physics output can be achieved with a relatively short run of the LAr1-ND experiment. In addition to the SBN physics program, LAr1-ND will have a development program serving as an engineering prototype for LArTPCs for long-baseline CP-violation searches in the future.

2. The LAr1-ND Measurement programme

While LAr1-ND, in conjunction with MicroBooNE and the ICARUS-T600, is a critical part of the oscillation physics program, the detector allows a large number of relevant physics results as a stand alone experiment. This section briefly discusses a selection of these measurements.

Figure 1. 3-detector system of the Short Baseline Neutrino programme at Fermilab.

2.1. *The MiniBooNE Low Energy Excess*

The MiniBooNE experiment [4] has observed an excess in low energy electron neutrino events, the nature of which has yet to be characterised. The main physics goal of MicroBooNE is to do so. Observation of a low energy excess signal by MicroBooNE in the years leading up to the beginning of LAr1-ND data taking would immediately lead to the question of whether that excess is intrinsic to the beam or appears over the 470 m distance between source and detector. LAr1-ND, at 110 m from the BNB target, can search for the same excess in a relatively short time.

It has been estimated [5] that for an exposure of 2.2×10^{20} protons on target (POT) a MiniBooNE-like excess would result in 803 excess ν_e charged current (CC) events, compared to a background of 3,177 events expected in the 200 - 650 MeV lepton energy range. Considering the statistical and systematic uncertainties, this is a 5.4σ signal.

2.2. *Neutrino Interaction Cross-section Measurements*

Neutrino-nucleus interactions are critical to understand in neutrino oscillation experiments, including the future liquid argon long-baseline program. LAr1-ND provides an ideal venue to conduct precision cross section measurements in the GeV energy range. The experiment will collect enormous neutrino event samples and, continuing the studies done by MicroBooNE and ICARUS, will make the world's highest statistics cross section measurements for many ν-Ar scattering processes.

In LAr1-ND more than 2 million neutrino interactions will be collected per year in the full active volume (assuming 2.2×10^{20} POT), with 1.5 million ν_μ and 12,000 ν_e charged current events, 250,000 neutral current events identifiable by a single nucleon recoil track, and a few hundred elastic scattering events. One year exposure of LAr1-ND will provide an event sample 6-7 times larger than will be available in the full MicroBooNE Phase I run.

In addition to the above measurements, we will be able to directly compare neutrino cross sections off carbon (A=12) and argon (A=40) targets to search for a nuclear dependence in cross section, since LAr1-ND will use the same neutrino beam used by the MiniBooNE experiment.

3. The LAr1-ND Detector

3.1. *The Cryostat and Infrastructure*

The LAr1-ND detector will make use of membrane cryostat technology as demonstrated by the 35t LBNE prototype [6]. The LAr1-ND cryostat will use the same commercial technology, using

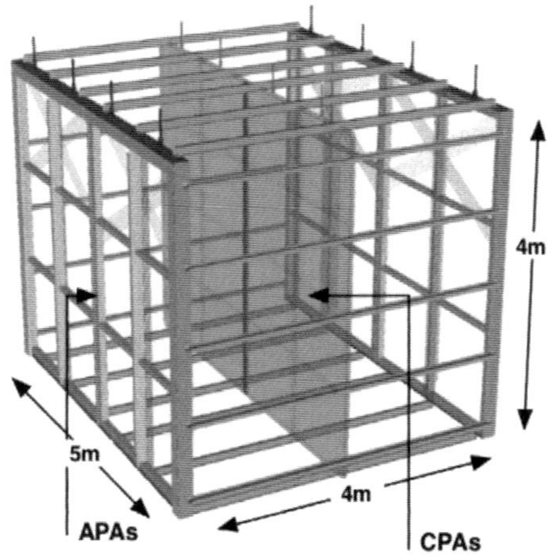

Figure 2. Conceptual drawing of the LAr1-ND TPC.

passive foam insulation, but with several key differences. The LAr pump will be located outside of the cryostat, allowing easier access. It is also planned to have a much smaller warm ullage than has been thus far demonstrated, which creates a less contaminated volume of gas, and hence a smaller scale purification system.

The detector system will be housed in a new building, located adjacent to the existing SciBooNE hall, as shown in figure 1.

3.2. TPC design

The TPC baseline design is a 5 m × 4 m × 4 m active volume, with two 2 m drift regions. The charge is drifted from a central cathode, to wire readout planes at either end of the detector. Both consist of multiple anode plane assemblies and cathode plane assemblies (APA and CPA respectively), as shown in figure 2.

Each APA has dimensions 2.5 m × 4 m, and consists of three wire planes; a vertical collection plane (Y), and two induction planes (U,V) at $\pm 60^\circ$ angles to the vertical, as illustrated in figure 3. The wire pitch is 3 mm, which is identical to that used in the T600 and MicroBooNE. This enables electron / photon separation with very similar efficiencies. Bias voltages of approximately -200 V, 0 V, and +500 V will be applied to the (U,V,Y) wire planes, respectively, to provide the 100% transparency condition necessary to allow all electrons to pass through the U and V planes and be collected by the Y plane.

All wires in the APAs are bonded mechanically with epoxy and terminated electrically with solder onto bonding boards made out of G10, which also provide connection to the readout electronics. The APA uses a similar wire bonding method to the one developed for the LBNF APAs, but without the continuous helical wrapping.

In order to minimize the cost of the readout electronics, each APA has cold readout electronics on two edges only. The U wires of each APA are electrically connected via flexible jumper connections at the adjoining edges, as shown in figure 3. The V wires are similarly connected.

In this design there is a gap of 15 mm between the two active apertures of the APAs which creates a dead readout region. To overcome this issue there is an option to insert a printed circuit board in the gap between the two APAs, with biased electrodes, such that the incoming electrons are deflected towards the active region of the wire frame, as shown in figure 4. This will eliminate electron loss, at the expense of distorted electric fields. The track reconstruction due

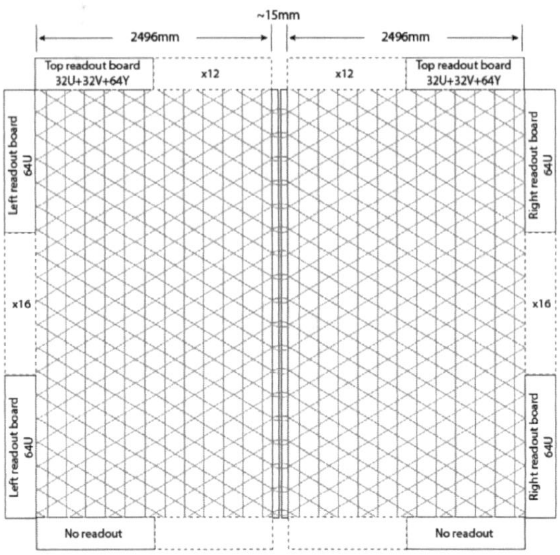

Figure 3. A schematic of the bridged APA concept.

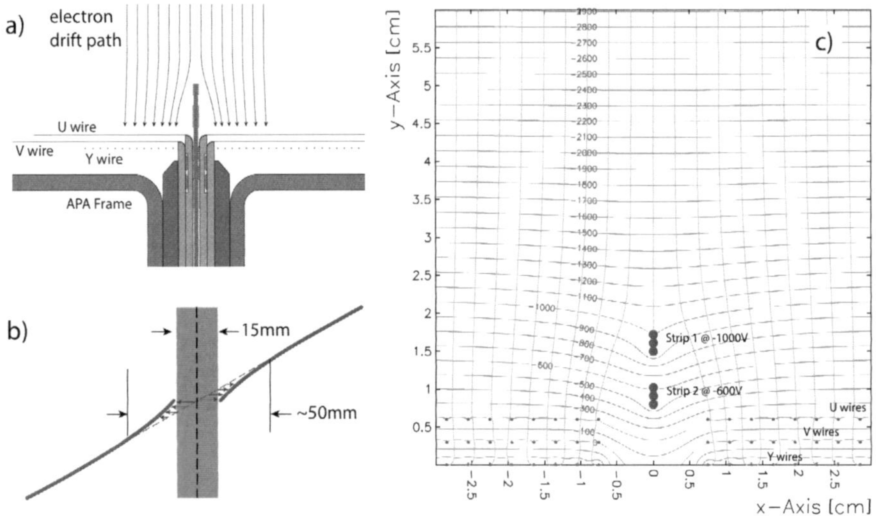

Figure 4. a) A schematic showing the biased-electrode inserted between the two APA frames, to divert electrons to the nearby active regions. b) The distortion on the reconstructed electron tracks. c) A Garfield simulation of the electron drift lines in a two strip configuration.

to the electric field distortions can be mapped out, and corrected at the reconstruction stage. This field shaping concept will be implemented at one section of the LBNE 35 ton TPC and evaluated during Phase II operation.

LAr1-ND aims to provide a verified alternative to the wrapped APA design, with a cost and space efficient design. This method of tiling the APAs and minimising dead space is extremely relevant R&D as we look at moving towards larger volume LAr TPCs.

The key design parameters of the LAr1-ND TPC are summarised in table 1.

Table 1. LAr1-ND TPC key design parameters

TPC parameter	**Value**
TPC active volume	112 metric ton
TPC dimensions	5 m (L) × 4 m (H) × 4 m (W)
Maximum drift time	1.28 ms, 2 m max. drift length
Anode Plane Assembly dimensions	2.5 m × 4 m
Wire Properties	150 μm diameter, CuBe
Wire planes	3 per APA, U and V at ±60° to vertical (Y)
Cathode bias	-100 kV at 500 V/cm drift field
Number of Wires	2816 channels/APA, 11264 wires total in TPC
Wire tension	0.5 kg at room temperature

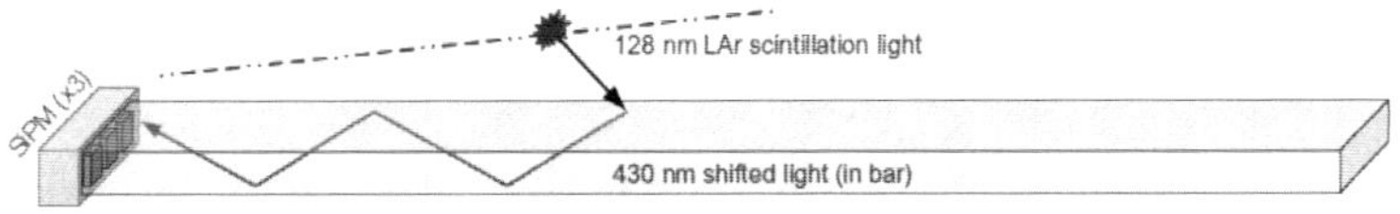

Figure 5. A wavelength shifting light guide bar, read out by SiPMs.

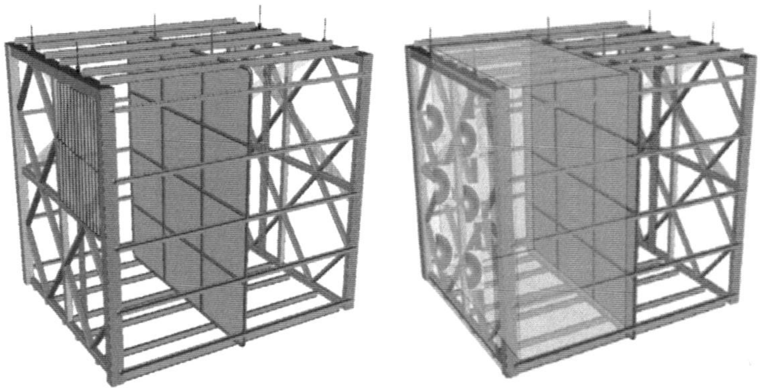

Figure 6. Two light collection system designs for the LAr1-ND TPC. Left: The light guide bar design. Right: TPB-coated reflector tile system.

3.3. Photon readout

LAr1-ND provides an excellent test-bed for light collection systems in a LAr detector. There are two proposed light collection systems, an overview of each is presented here. The opportunity to test new approaches, possibly side-by-side is an important one for the goal of informing an optimized design for use in future LAr detectors for long-baseline neutrino physics.

One of the approaches is very similar to the current baseline design for LBNF. The light collection system consists of acrylic bars mounted behind the APA frame, as shown in figure 6. These acrylic bars are dip-coated in Tetraphenyl butadeine (TPB) wavelength shifter which converts the incident primary scintillation light with wavelength 128 nm to 430 nm light, which can be efficiently read out (see figure 5). Each bar is read out by 3 Silicon Photomultipliers (SiPM), giving position resolution of 2.5 cm in the vertical direction.

Another approach being considered is based on a concept adapted from liquid argon dark matter detectors. In this design, the inner surfaces of the detector volume are covered with TPB-coated reflector foils, in order to enhance light collection. The motivation behind this is to naturally enhance the light yield without need for additional readout channels. The reflector

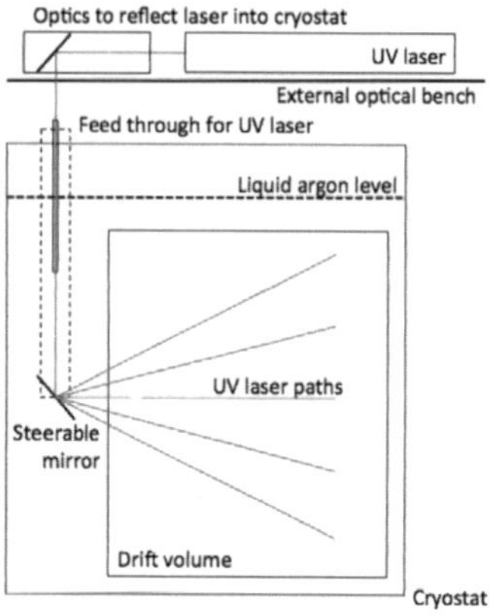

Figure 7. A schematic of a UV laser system producing a fan of straight tracks from one laser source in a LArTPC.

foils proposed are a modular system of tiles lining the inner field cage,and the light is collected at the APA planes (as shown in figure 6) by high quantum efficiency cryogenic PMTs, SiPMs, or a combination of the two. A system similar to this has been implemented in the LArIAT experiment [7].

The use of TPB-coated cryogenic PMTs is also being considered, analogous to the system implemented in the ICARUS and MicroBooNE detectors.

3.4. Detector drift field calibration system

The knowledge of the electric field inside the drift volume of a TPC is a key aspect for performing subsequent event reconstruction. Since distortions of particle tracks due to field non-uniformities are indistinguishable from particle multiple scattering, they affect the accuracy of the particle momentum reconstruction based on track scattering angles.

Deviations of the field map from perfect uniformity in a LArTPC may arise due to accumulation of positive argon ions in the drift volume. Since the LAr1-ND detector sits at the surface, there will be an estimated 2200 cosmic ray muons per second in the fiducial volume of the detector. This, in addition to the ions produced in neutrino interactions leads to space-charge; an accumulation of low mobility positive ions in the detector volume. This causes electric field non-uniformities, which change as a function of time.

In order to measure these electric field non-uniformities, a pulsed UV laser system will be implemented in the detector. This system, as described in [8] will produce straight ionisation tracks at defined locations in the liquid argon, using a $\lambda = 266\,\text{nm}$ source to produce ionisation via multi-photon absorption. The laser tracks are created one by one using pulsed laser beams and a steerable mirror, as shown in figure 7. The reconstructed position of the tracks can then be mapped to the known location of the ionisation deposition, and the curvature of the tracks used to map the distortions of the electric fields.

This system was developed in ARGONTUBE [9], and a similar system has been implemented in the MicroBooNE detector.

3.5. Electronics

The aim of the TPC readout is to digitize, compress losslessly and record the TPC signals upon the reception of a variety of triggers. The electronics readout of the 11,264 APA wire channels consists of three parts: cold electronics, warm interface electronics, and signal feedthrough. The system will be built upon the technology developed for use in both the MicroBooNE and

Journal of Physics: Conference Series **650** (2015) 012007 doi:10.1088/1742-6596/650/1/012007

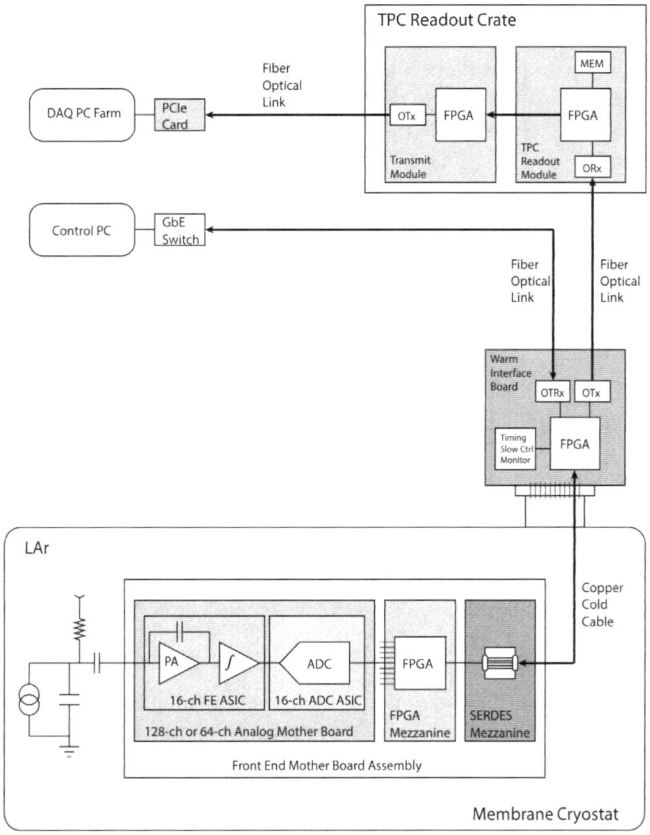

Figure 8. Block diagram showing the electronics for TPC readout.

LBNE 35t prototype, however, the major difference is the intention to put all of the front-end electronics inside the cryostat, operating in the cold.

A block diagram showing the readout of a single APA channel is shown in figure 8. The signal from each wire is pre-amplified and shaped by a CMOS analog front end ASIC, then digitized by a CMOS ADC ASIC inside the cryostat. The digitized signal is sent to an FPGA, which aggregates data from multiple ADC chips and multiplexes it to high speed serial links. The serial data is then sent over cold cables through a feed-through to the warm interface board installed outside of the cryostat on top of the signal feed-through. The warm interface board receives the electrical serial data from the cold electronics and converts it to optical signals for transmission over a fiber optical link to the TPC readout module, housed in a crate. Once the signal arrives at the TPC readout module, it is processed in an FPGA for compression, reduction, and storage. Processed data is buffered on board temporarily and then transmitted to DAQ PCs through the crate backplane and optical links. Data received on PCs is then stored in hard drives for further analyses.

4. Summary and Timeline

LAr1-ND is the new LArTPC in the SBN program. It not only has an exciting physics program, but also provides a great opportunity for detector development towards future large liquid argon detectors for long-baseline neutrino experiments.

Currently, there is a large amount of R&D activity towards realising the detector. The preliminary design is complete, with a conceptual design review (CDR) submitted in January 2015, and a technical design report to be submitted mid-2015. Ground breaking for the detector building is planned to take place in August 2015, with the cryostat installation in November 2016. The detector installation will take place in late 2017, with a view to a first cooldown in November 2017, and first beam data in April 2018.

The proposed schedule for the LAr1-ND experiment is a very aggressive one, with design, fabrication, assembly and installation of the detector before the end of 2017. Nevertheless, the collaboration is building on the experience of the design, construction, and operation of existing liquid argon TPCs, and this gives the project the potential to move forward quickly, attaining these goals.

5. References

[1] Accarri R *et al.* 2013 (*Preprint* 1309.7987)
[2] MicroBooNE project 2012 The MicroBooNE Technical Design Report CD3b review
[3] Rubbia C, Antonello M, Aprili P, Baibussinov B, Ceolin M B *et al.* 2011 *JINST* **6** P07011 (*Preprint* 1106.0975)
[4] Aguilar-Arevalo A *et al.* (MiniBooNE Collaboration) 2009 *Phys.Rev.Lett.* **102** 101802 (*Preprint* 0812.2243)
[5] Adams C *et al.* 2015 (*Preprint* 1503.01520)
[6] Montanari D *et al.* (LBNE) 2013 *FERMILAB-CONF-13-076-PPD*
[7] Szelc A (LArIAT) 2013 *JINST* **8** C09011
[8] A Ereditato *et al* 2014 (*Preprint* 1406.6400)
[9] Ereditato A, Hsu C, Janos S, Kreslo I, Messina M *et al.* 2013 *JINST* **8** P07002 (*Preprint* 1304.6961)

Low Background Micromegas in CAST

J G Garza[1], S Aune[2], F Aznar[1,7], D Calvet[2], J F Castel[1], F E Christensen[4], T Dafni[1], M Davenport[5], T Decker[3], E Ferrer-Ribas[2], J Galán[2], J A García[1], I Giomataris[2], R M Hill[3], F J Iguaz[1], I G Irastorza[1], A C Jakobsen[4], D Jourde[2], H Mirallas[1], I Ortega[1,8], T Papaevangelou[2], M J Pivovaroff[3], J Ruz[3], A Tomás[1,9], T Vafeiadis[5], J K Vogel[3]

[1] Grupo de Física Nuclear y Astropartículas, University of Zaragoza, Zaragoza, Spain
[2] Irfu, CEA, Centre de Saclay, Gif sur Yvette, France
[3] Lawrence Livermore National Laboratory, Livermore, CA, USA
[4] Technical University of Denmark, DTU Space Kgs. Lyngby, Denmark
[5] CERN, European Organization for Particle Physics and Nuclear Research, Geneva, Switzerland

E-mail: `jgraciag@unizar.es`

Abstract. Solar axions could be converted into x-rays inside the strong magnetic field of an axion helioscope, triggering the detection of this elusive particle. Low background x-ray detectors are an essential component for the sensitivity of these searches. We report on the latest developments of the Micromegas detectors for the CERN Axion Solar Telescope (CAST), including technological pathfinder activities for the future International Axion Observatory (IAXO). The use of low background techniques and the application of discrimination algorithms based on the high granularity of the readout have led to background levels below 10^{-6} counts/keV/cm^2/s, more than a factor 100 lower than the first generation of Micromegas detectors. The best levels achieved at the Canfranc Underground Laboratory (LSC) are as low as 10^{-7} counts/keV/cm^2/s, showing good prospects for the application of this technology in IAXO. The current background model, based on underground and surface measurements, is presented, as well as the strategies to further reduce the background level. Finally, we will describe the R&D paths to achieve sub-keV energy thresholds, which could broaden the physics case of axion helioscopes.

1. Micromegas for axion searches

Axions were proposed to solve the strong-CP problem more than 35 years ago by R. D. Peccei and H. R. Quinn[1]. Now, they are still the most compelling solution, and one of the few viable candidates to compose the dark matter. Axion helioscopes [2, 3] aim to detect solar axions by their conversion into x-rays (1-10 keV) in the presence of strong magnetic fields. The CERN Axion Solar Telescope (CAST)[4, 5, 6] is the most powerful implementation of this technique so far, setting the most stringent limits on the axion-photon coupling constant $g_{a\gamma}$ for a wide range

[7] Present address: Centro Universitario de la Defensa, Universidad de Zaragoza, Ctra. de Huesca s/n, 50090 Zaragoza, Spain
[8] Present address: CERN, European Organization for Particle Physics and Nuclear Research
[9] Present address: High Energy Physics group, Brackett Laboratory, Imperial College, London, UK

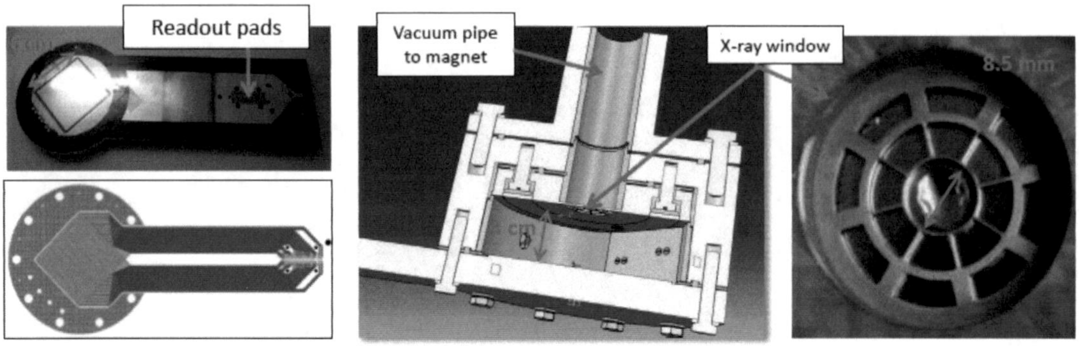

Figure 1. The CAST microbulk Micromegas readout parallelizes 6×6 cm in 120×120 strips (left). Detector sketch showing the main components of the detection chamber (center). The x-ray window is a gas-tight, 4 μm, aluminized mylar foil glued on a spider-web patterned copper strong-back that withstands the differential pressure to the magnet vacuum system (right).

of axion masses [5, 6]. The future International Axion Observatory (IAXO) aims to increase the sensitivity to the axion-photon coupling constant by 1-2 orders of magnitude [7, 8].

The sensitivity of an axion helioscope can be considered in terms of four main ingredients (magnet, optic, x-ray detector, exposure) of such an experiment as $g_{a\gamma} \propto b^{1/2}\epsilon^{-1} \times s^{1/2}\epsilon_o^{-1} \times (BL)^{-2}A^{-1} \times t^{-1/2}$ [9]. Here, t is the exposure time, L and A are the length and cross-sectional area of the magnet, B the intensity of the magnetic field, s and ϵ_o are the focusing spot area and efficiency of the x-ray optics, and b and ϵ are respectively the background level and signal efficiency of the x-ray detectors.

Therefore, high-efficient low-background x-ray detectors are mandatory for an axion helioscope. Microbulk Micromegas (MM) x-ray gaseous detectors [10] are currently used to equip three of the four CAST magnet bores, and they have been proposed as the baseline detector technology for IAXO (figure 1). The choice of this technology is based on several properties: a) the readout is highly granular, allowing a strong background suppression based on the different topology of the x-ray signal (point-like) and the gamma and muon background (extended, non-symmetric tracks) [11]; b) all the detector components are carefully chosen to have low radioactivity, including the Micromegas readout, whose radioactivity has been measured in a high-purity germanium detector in the Canfranc Underground Laboratory (LSC) [12]. c) active and passive shielding techniques can be applied as in rare event searches; d) their manufacture relies on a consolidated technique that guarantees the stability of the readout over long running periods; high and uniform gains are achieved, and good energy resolution (12% FWHM at 5.9 keV) can be reached [13, 14].

The integration of an x-ray focusing device with a low background MM detector is presented in section 2. The latest background levels achieved in CAST-MM detectors [15] and in the Canfranc Underground Laboratory make the MM technology an excellent candidate for accomplishing the IAXO requirements, namely $10^{-7} - 10^{-8}$ keV^{-1}cm^{-2}s^{-1} in the energy range of interest (RoI). The current background model and the strategies to further reduce the background level are presented in section 3, while in section 4 some research paths for lowering the energy threshold are presented.

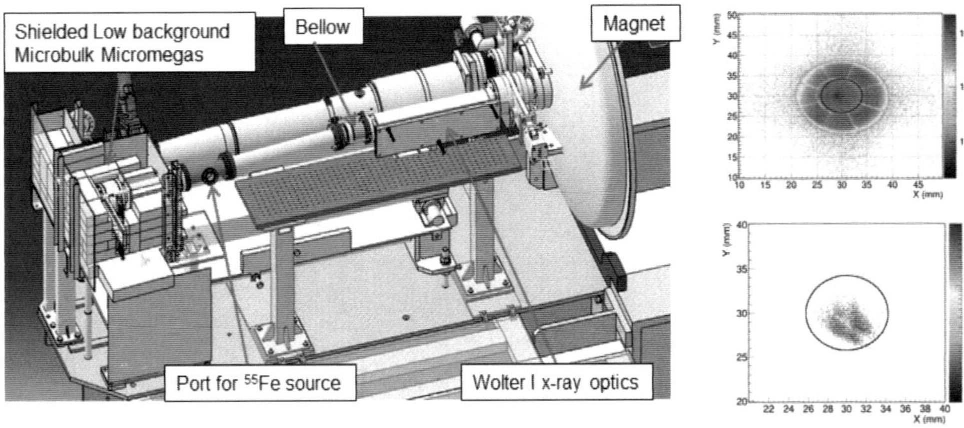

Figure 2. Sketch of the CAST detection line composed of a low background MM placed at the focal point of an x-ray telescope (left). Intensity map of a regular ^{55}Fe calibration (right-up) where the shadow of the strong-back is observed. The same is represented for x-rays coming through the optics from a source placed 13 m away (right-down).

2. MM and x-ray optics

The helioscope technique can be enhanced by the use of an x-ray focusing device, increasing the signal-to-background ratio and thus the sensitivity. A new dedicated x-ray optic was installed at one end of the CAST magnet in 2014, with a low background MM at its focal plane. On top of increasing CAST's sensitivity, it is a demonstration of the techniques proposed in the conceptual design report for IAXO [8]. The line is composed of a ~5 cm diameter, 1.3 m focal-length Wolter I x-ray telescope (XRT) and a shielded MM detector made of radiopure components (figure 2, left). For the first time, the telescope has been specifically designed for axion research using NUSTAR technology [16, 17].

This line was successfully commissioned shortly after its installation at CAST, and it is currently looking for solar axions. Figure 2 (top-right) shows the intensity map registered by the MM detector produced with a regular ^{55}Fe source placed behind the x-ray optic. The intensity map produced in the MM detector by an x-ray source focused by the telescope is shown in figure 2 (bottom-right): it approximately defines the region where the axion signal is expected. The increase in sensitivity due to the x-ray optic focusing is roughly estimated as a factor $5-8$. A detailed paper reporting the design, operation and results of the MM+XRT system is in preparation, as well as a more technical paper on the x-ray focusing device.

3. CAST-MM background model and strategies for reduction

The application of background suppression techniques, the reliability of the microbulk technology, the upgrade of the readout electronics and the tuning of the rejection algorithms led to a level reduction of more than two orders of magnitude in CAST-MM detectors over the last ten years (figure 3) [11, 18]. The current background level is below 10^{-6} keV^{-1}cm^{-2}s^{-1} in the CAST RoI [15]. The corresponding background energy spectrum is characterized by a fluorescence peak at 8 keV (from copper K_α emission), its escape peak at 5 keV and by the argon K_α emission line at 3 keV (figure 3). The operation of a replica of the detector in the LSC sets a level as low as ~10^{-7} keV^{-1}cm^{-2}s^{-1}, almost at the required IAXO levels.

Further background reductions depend on the identification of the source of the remaining events and on the application of the necessary techniques to minimize them. The current

Journal of Physics: Conference Series **650** (2015) 012008 doi:10.1088/1742-6596/650/1/012008

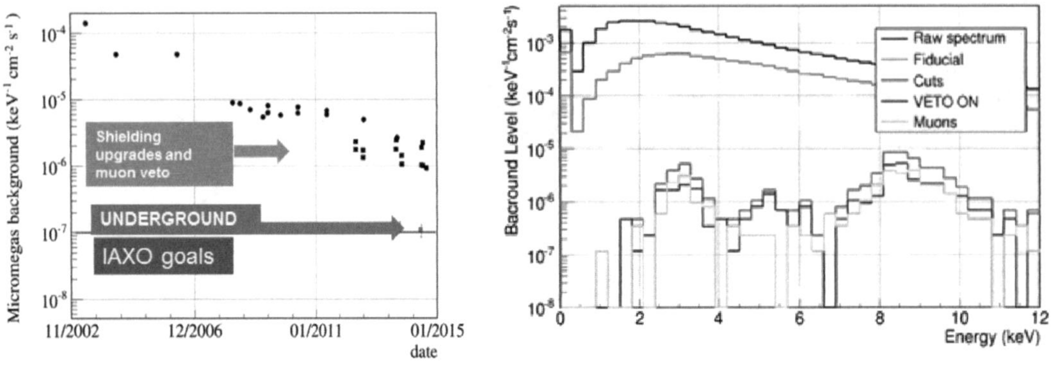

Figure 3. Evolution of the background level in CAST, the limit obtained in the LSC and IAXO goals (left). Background energy spectrum of the CAST line in 2014 (right).

Contribution	Level ($keV^{-1}cm^{-2}s^{-1}$) Before	After	Reduction technique applied
Gamma flux	7×10^{-5}	Negligible?	Full coverage by 10 cm lead shielding
Radon	8×10^{-7}	Negligible	Nitrogen flux inside the shielding
Cosmic muons	2×10^{-6}	7×10^{-7}	95% coverage by an active muon veto
Al cathode	5×10^{-7}	Negligible	Replacement by an ultrapure copper cathode
LSC limit	1.1×10^{-7}		^{39}Ar?, Neutrons?, cosmic activation?

Table 1. Summary of the different contributions to the background level in the RoI of the CAST-MM detectors and the techniques applied to reduce them (table extracted from [20]). A detailed description of the measurements and their associated error is given in [11].

understanding of the background sources is summarized in table 3. This model is based on in-situ measurements at CAST, tests underground (at the LSC) and at surface, as well as in Geant4 simulations. While the contributions from gamma rays, radon and internal radioactivity have been reduced to negligible levels, the dominant component is caused by cosmic muons and their secondary products, generated after their interaction in the setup components. The origin of the events limiting the LSC performance is unknown. Some hypotheses are the β-decay of ^{39}Ar present in the detection medium, neutrons or cosmic activation of the materials.

The strategy to reduce background at surface level are based on the installation of thicker and more compact shielding and in a 4π enlarged muon veto system. Pushing the lowest underground limit requires a change in the active gas to xenon or neon, or the installation of a neutron shielding. These activities are being developed in the context of the R&D phase for the IAXO technical design report.

4. Towards lower energy thresholds

The efficiency of the x-ray focusing device shifts the signal to lower energies. Even in the hadronic axion production mechanism, an important fraction of the signal is at sub-keV energies. Moreover, axions could be produced at the Sun via non-hadronic processes at an energy peaking around 1 keV [19]. More exotic searches, like paraphotons or chameleons also peak at very low energies. These facts motivate the use of sub-keV energy threshold detectors.

The R&D paths for lowering the energy threshold include: the use of more transparent x-ray windows made out of other materials or different geometries than presently used; other gas mixtures with higher gain at the operation point [14]; self-triggered electronic readouts to improve the signal-to-noise ratio [21]; calibration at lower energies that will provide new analysis tools based on the dependence of the event topology on the x-ray energy (see [20] for more details).

5. Conclusions

The Micromegas detectors in CAST have reduced their background level by more than two orders of magnitude in the last decade, reaching values below 10^{-6} keV^{-1}cm^{-2}s^{-1}. The different measurements performed at CAST and at the LSC have allowed to build a background model with our understanding of the system. The main contributions have been identified and partially or totally reduced. We have presented the open R&D paths for further reducing the remaining contributions together with new research paths for lowering the energy threshold of the MM detectors. Finally, the best levels achieved in the measurements performed in the LSC motivate the use of this technology in IAXO.

Acknowledgments

We want to thank our colleagues of CAST for many years of collaborative work in the experiment, and many helpful discussions and encouragement. Part of this work was performed under the auspices of the U.S. Department of Energy by Lawrence Livermore National Laboratory under Contract DE-AC52-07NA27344. We thank R. de Oliveira and his team at CERN for the manufacturing of the microbulk readouts. We also thank the LSC staff for their help in the support of the Micromegas setup at the LSC. The authors would like to acknowledge the use of Servicio General de Apoyo a la Investigación-SAI, Universidad de Zaragoza. F. I. acknowledges the support from the *Juan de la Cierva* program of the MINECO. We acknowledge support from the European Commission under the European Research Council T-REX Starting Grant ref. ERC-2009-StG-240054 of the IDEAS program of the 7th EU Framework Program. We also acknowledge support from the Spanish Ministry of Economy and Competitiveness (MINECO) under contracts ref. FPA2008-03456, FPA2011-24058 and EIC-CERN-2011-0006, as well as under the CPAN project ref. CSD2007-00042 from the Consolider-Ingenio 2010 program. Part of these grants are funded by the European Regional Development Fund (ERDF/FEDER).

References

[1] Peccei R D and Quinn H R 1977 *Rev. Lett.* **38** 1440-1443
[2] Sikivie P 1983 *Phys. Rev. Lett* **51** 1415
[3] van Bibber K *et al.* 1989 *Phys. Rev.* **D39**
[4] CAST Collaboration, Zioutas K *et al.* 2005 *Phys. Rev. Lett.* **94** 121301
[5] CAST Collaboration, Adriamonje S 2007 *et al.JCAP* **04** 010
[6] CAST Collaboration, Arik E 2014 *et al.Phys. Rev. Lett.* **112** 091302
[7] Irastorza I G *et al.* 2013 *CERN-SPSC-2013-022, SPSC-I-242*
[8] Armengaud E *et al.* 2014 *JINST* **9** T05002
[9] Irastorza I G *et al.* 2011 *JCAP* **1106** 013
[10] Adriamonje S *et al.* 2010 *JINST* **5** P02001
[11] Aunse S *et al.* 2014 *JINST* **9** P01001
[12] Cebrian S *et al.* 2011 *Astropart. Phys.* **34** 354
[13] Cebrian S *et al.* 2010 *JCAP* **1010** 010
[14] Iguaz F J *et al.* 2012 *JINST* **7** P04007
[15] Garza J G *et al.* 2013 *JINST* **8** C12042
[16] Harrison F A *et al.* 2013 *Astrophysical Journal* **770** 103
[17] Jakobsen A C *et al.* 2013 *Proc. SPIE* **8861**
[18] Garcia J G *et al.* 2013 *J. Phys Conf Ser.* **460** 012003
[19] Redondo J 2013 *JCAP* **1312** 008

[20] Iguaz F J *et al.* 2014 *PoS TIPP2014* 295
[21] Anvar S *et al.* 2011 *Proc. IEEE Nuclear Science Symp.* pp. 745-749

Journal of Physics: Conference Series **650** (2015) 012009 doi:10.1088/1742-6596/650/1/012009

The IAXO Helioscope

E Ferrer Ribas[1], E Armengaud[1], F T Avignone[2], M Betz[3], P Brax[4], P Brun[1], G Cantatore[5], J M. Carmona[6], G P Carosi[7], F Caspers[3], S Caspi[8], S A Cetin[9], D Chelouche[10], F E Christensen[11], A Dael[1], T Dafni[6] M Davenport[3], A V Derbin[12], K Desch[13], A Diago[6], B Döbrich[14], I Dratchnev[12], A Dudarev[3], C Eleftheriadis[15], G Fanourakis[16], J Galán[6], J A García[6], J G Garza[6], T Geralis[16], B Gimeno[17], I Giomataris[1], S Gninenko[18], H Gómez[6], D González-Díaz[6], E Guendelman[19], C J. Hailey[20], T Hiramatsu[21], D H H. Hoffmann[22], D Horns[23], F J Iguaz[6], I G Irastorza[6], J Isern[24], K Imai[25], J Jaeckel[26], A C Jakobsen[11], K Jakovčić[27], J Kaminski[13], M Kawasaki[28], M Karuza[29], M Krčmar[27], K Kousouris[3], C Krieger[13], B Lakić[27], O Limousin[1], A Lindner[14], A Liolios[15], G Luzón[6], S Matsuki[30], V N Muratova[12], C Nones[1], I Ortega[6], T Papaevangelou[1], M J Pivovaroff[7], G Raffelt[31], J Redondo[6], A Ringwald[14], S Russenschuck[3], J Ruz[7], K Saikawa[30], I Savvidis[15], T Sekiguchi[28], Y K Semertzidis[33], I Shilon[6,3], P Sikivie[34], H Silva[3], H H J ten Kate[3], A Tomas[6], S Troitsky[18], T Vafeiadis[3], K van Bibber[35], P Vedrine[1], J A Villar[6], J K Vogel[7], L Walckiers[3], A Weltman[36], W Wester[37], S C. Yildiz[9], K Zioutas[3,38]

[1]CEA, Irfu, Centre de Saclay, Gif sur Yvette, France
[2]Physics Department, University of South Caroline, Columbia, SC, USA
[3]CERN, Genève, Switzerland
[4]CEA, IPHT, Centre de Saclay, Gif sur Yvette, France
[5]INFN, Sezione di Trieste & Università di Trieste, Italy
[6]Lab. de Física Nuclear y Altas Energías, Univ. de Zaragoza, Zaragoza, Spain
[7]Physical and Life Sciences Directorate, LLNL, Livermore, California, USA
[8] LBNL, Berkeley, California, USA
[9] Dogus Univ., Istanbul, Turkey
[10]Physics Department, Univ. of Haifa, Haifa, 31905 Israel
[11]DTU Space, Tech. Univ. of Denmark, Copenhagen, Denmark
[12]St.Petersburg Nuclear Physics Institute, St.Petersburg, Russia
[13]Physikalisches Institut der Universität, Bonn, Germany
[14]DESY, Hamburg, Germany
[15]Aristotle Univ. of Thessaloniki, Thessaloniki, Greece
[16]NCSR Demokritos, Athens, Greece
[17]Inst. de Ciencias de las Materiales, Univ. de Valencia, Valencia, Spain
[18]INR, Russian Academy of Sciences, Moscow, Russia
[19]Physics dep., Ben Gurion Univ., Beer Sheva, Israel
[20]Columbia Univ. Astrophysics Laboratory, New York, NY, USA
[21]Yukawa Inst. for Theoretical Physics, Kyoto Univ., Kyoto, Japan
[22]Technische Universität Darmstadt, IKP, Darmstadt, Germany
[23]Inst. für Experimentalphysik, Universität Hamburg, Hamburg, Germany
[24]Inst. de Ciències de l'Espai (CSIC-IEEC), Facultat de Ciències, Bellaterra, Spain

[25]Advanced Science Research Center, Japan Atomic Energy Agency, Tokai-mura, Ibaraki-ken, Japan
[26]Inst. für theoretische Physik, Universität Heidelberg, Philosophenweg 16, 69120 Heidelberg, Germany
[27]Rudjer Bošković Institute, Zagreb, Croatia
[28]Inst. for Cosmic Ray Research, Univ. of Tokyo, Tokyo, Japan
[29]Physics Dep. & Center for Micro and Nano Sciences and Technologies, Univ. of Rijeka, Croatia.
[30]Research Center for Low Temperature & Materials Sciences, Kyoto Univ., Kyoto, 606-8502 Japan
[31]Max-Planck-Institut für Physik, Munich, Germany
[32]Dep. of Physics, Tokyo Institute of Technology, Tokyo, Japan
[33]Center of Axion & Precision Physics, IBS, KAIST, Daejeon, Korea
[34]Department of Physics, Univ. of Florida, Gainesville, FL 32611, USA
[35]Dep. of Nuclear Engineering, Univ. of California Berkeley, Berkeley, CA, USA
[36]Univ. of Cape Town, South Africa
[37]Fermi National Accelerator Laboratory, Batavia, IL, USA
[38]Physics Department, Univ. of Patras, Patras, Greece

E-mail: esther.ferrer-ribas@cea.fr

Abstract.
The IAXO (International Axion Experiment) is a fourth generation helioscope with a sensitivity, in terms of detectable signal counts, at least 10^4 better than CAST phase-I, resulting in sensitivity on $g_{a\gamma}$ one order of magnitude better. To achieve this performance IAXO will count on a 8-coil toroidal magnet with 60 cm diameter bores and equipped with X-ray focusing optics into 0.20 cm^2 spots coupled to ultra-low background Micromegas X-ray detectors. The magnet will be on a platform that will allow solar tracking for 12 hours per day. The next short term objectives are to prepare a Technical Design Report and to construct the first prototypes of the hardware main ingredients: demonstration coil, X-ray optics and low background detector while refining the physics case and studying the feasibility studies for Dark Matter axions.

1. Introduction

The Standard Model (SM) of particle physics is a very successful theory that explains with a high level of accuracy the fundamental constituents of matter and their interactions, especially since the Higgs Boson was discovered at the Large Hadron Collider (LHC) in 2012.

Yet the SM is not a complete theory as there are still a number of unanswered questions among them the origin of Dark Matter (DM) and Dark Energy (DE), the matter-antimatter asymmetry of the universe, and why gravity can not be included in this theory...The search of signs of physics beyond the Standard Model at high energy will continue with the LHC run II from summer 2015 to 2018 at 13 TeV. However in a large number of more general theories, embedding the SM, low mass and very weakly interacting particles, the so called WISPs (Weakly Interacting Slim Particles) are predicted. Axions and axion-like particles (ALPs) are salient examples of WISPs. Axions are light pseudoscalar particles that arise in the context of the Peccei-Quinn[1] solution to the strong CP problem and can be Dark Matter candidates[2]. The Primakoff effect which allows the conversion of axions into photons in the presence of a strong magnetic field is used in all experimental searches. In helioscopes, the Sun as source of axions, can be scrutinised by directing a strong dipole magnet towards it and searching for the axion-photon conversion in the keV range. The CAST (Cern Axion Solar Telescope) experiment has been using a decommissioned LHC dipole magnet to convert solar axions into detectable x-ray photons since 2003. The CAST experiment data taking has provided the most restrictive experimental limits on the axion-photon coupling for a broad range of axion masses[3, 4, 5, 6, 7]. In order to reach further sensitivities a new axion helioscope, IAXO (International AXion Observatory)[9], is being designed and is described in the following sections.

2. Experimental set-up

The aim of IAXO is to improve the sensitivity in terms of axion-photon coupling, $g_{a\gamma}$, by a factor of ~ 20 with respect to the CAST experiment, meaning about 4-5 orders of magnitude in terms of signal intensity. The key ingredients are: a large and strong magnet, focusing optics and low background detectors. In order to justify the choices made for the different systems i.e. the magnet, the X-ray focusing optics and the X-ray detectors, the dependence of the sensitivity of IAXO with respect to each system is now discussed.

The discovery potential of the experiment depends on $N_\gamma/\sqrt{N_b}$ where the number of axion signal counts N_γ and background counts, N_b can be written as follows [9]:

$$N_\gamma \sim N^* \times g_{a\gamma}^4 = B^2 L^2 A \epsilon t \times g_{a\gamma}^4 \tag{1}$$

and

$$N_b = b a \epsilon_t t \tag{2}$$

where B, L and A concern the magnetic field, the length and the cross sectional area respectively. ϵ is the total efficiency $\epsilon = \epsilon_d \epsilon_o \epsilon_t$ taking into account the detector efficiency, ϵ_d, the optics throughput ϵ_o and the data taking efficiency ϵ_t, which is the amount of time where the magnet is following the sun. The background of the detector is denoted by b and given in counts $\mathrm{keV}^{-1}\,\mathrm{cm}^{-2}\mathrm{s}^{-1}$, a is the total focusing spot area and t is the duration of the data taking. We can define a figure of merit that takes into account the contribution of the different systems to the total sensitivity:

$$f = \frac{N_\gamma}{\sqrt{N_b}} = f_M f_{DO} f_t \tag{3}$$

with

$$f_M = B^2 L^2 A; \; f_{DO} = \frac{\epsilon_d \epsilon_o}{\sqrt{ba}}; \; f_t = \sqrt{\epsilon_t t}. \tag{4}$$

These relations assume that the axion-photon conversion is fully coherent throughout the length of the magnet L and that background counts are in the Gaussian regime. From these expressions, one can see that the sensitivity of a helioscope is mainly driven by the magnet parameters. In order to surpass CAST performances, it does not seem feasible to build a dipole that would outperform the CAST LHC dipole parameters in a significant manner. Instead the toroidal configuration might enable to keep a BL close to the levels achieved by CAST while increasing significantly the cross sectional area. Moreover it is desirable to have X-ray focusing optics in most of the magnet bores with an entrance that fits the magnet bore and with a large throughput ϵ_o and a small spot area a. The background level of X-ray detectors should be 10^{-7} counts $\mathrm{keV}^{-1}\,\mathrm{cm}^{-2}\mathrm{s}^{-1}$ or lower. With the present performance of the Micromegas detectors this should not be a problem especially after the 2014 data taking during which a new state-of-the-art microbulk detector was tested. The tracking efficiency should also be improved with a platform that allows tracking the sun for at least 30 or 50% of the day.

With these considerations and after a preliminary study, a conceptual design has been proposed with an 8-coil toroidal magnet with 60 cm diameter bores equipped with optics focusing X-rays into 0.20 cm^2 spots coupled to ultra-low background Micromegas X-ray detectors. The magnet would be on a platform that would allow solar tracking for 12 hours per day. A sketch of the conceptual design is given in figure 1 as well as a possible implementation.

The concept of a new generation of helioscope was proposed in [9] with ultra-low background Micromegas detectors as the baseline option for the X-ray detectors. A letter of intent was proposed to CERN [11] and the conceptual design is described in [12]. In the next sections we

Journal of Physics: Conference Series **650** (2015) 012009 doi:10.1088/1742-6596/650/1/012009

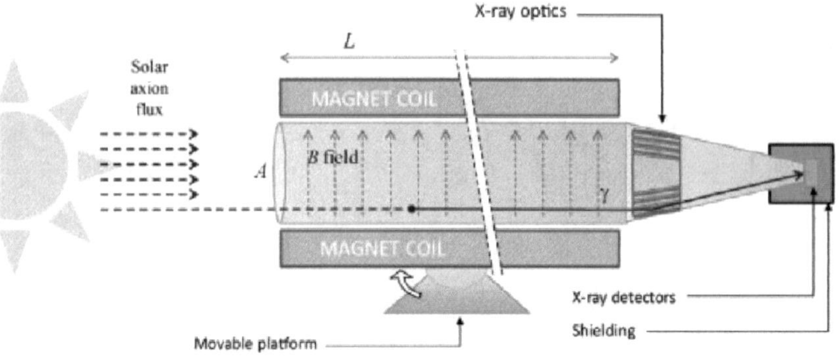

Figure 1. Sketch of the principle of IAXO.

will describe shortly the main ingredients: the magnet, the low background detectors and the focusing optics. The expected sensitivity will be given in section 3.

2.1. Magnet

After a prospective study of the available magnets on the various accelerators and experiments, the CERN magnet group concluded that a new magnet is required in order to achieve the required sensitivity. A well-known proven engineering technology has been chosen, the NbTi technology. With this technology, the attainable peak magnetic field is 5-6 T. Therefore, in order to reach the required magnet FOM $f_m = L^2B^2A$, the aperture is the parameter that can be enlarged easily.

The first optimisation of the geometry of the IAXO superconducting magnet based on simulations has shown that the toroidal geometry, inspired by the ATLAS barrel and end-cap toroids, is the preferred one [9]. These simulations have been realised within the CERN magnet group. The study has shown the optimal configuration is a toroid with 8 magnet bores each 1 m wide and 25 m in length reaching a peak magnetic field of 5.4 T and an averaged field in the bores of 2.5 T (critical magnetic field 8.8 T). The study of the magnet takes into account the implementation options of the other of the systems. For instance, the optimisation of the different parameters will not be the same depending on whether the optics are positioned as close as possible to the inner radius of the the toroid ("field dominated") or between each pair of racetrack coils ("area dominated"). The study also shows that is better to have an increased fraction of the aperture of the optics exposed to X-rays favouring the "area dominated" options as shown in figure 2.

The toroidal magnet consists of eight coils and their casing within an inner cylindrical support for the magnetic forces. The operational current is 12.3 kA with a stored energy of ~500 MJ. Safety margins have been taken into account in the operational current and temperature to allow proper operation. The operational temperature will be 4.5 K. For safety it is also necessary to design a quench protection system to allow the safe release of the stored energy.

The magnet needs to follow the sun as long as possible requiring a horizontal and vertical movement. In order to track for a half day a horizontal movement of 360° will be required with a vertical movement of 25°. The rotation system can be seen in figure 1: the whole system is supported by the central post with an inclination system that allows the vertical movement sitting on a rotating platform allowing the full horizontal rotation.

In order to validate the design of the IAXO magnet, it is envisaged to build and test a single short (2 m) prototype coil in the short term to demonstrate the feasibility of the IAXO toroid

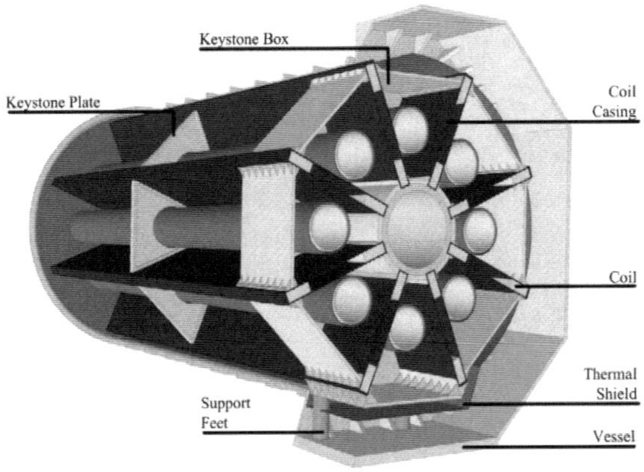

Figure 2. Transverse section of the magnet cryostat, showing the cold mass and its supports, surrounded by a thermal shield and the vacuum vessel.

magnet.

2.2. Focusing X-ray optics
The use of X-ray telescopes allows IAXO to enhance the signal to background ratio by concentrating the potential signal in a small spot. The main requirements of helioscope optics are high efficiency in the 1–10 keV range and an aperture that fits the area of the magnet bore with a solid angle acceptance that is greater than the 0.3 arcmin (the extent of the solar core where axions are produced). Taking into account these requirements, using reflective X-ray optics seems to be the most appropriate choice.

The parameters of the optics need to be optimised in order to have the smallest spot, a, and the highest reflectivity. To have a small spot, the optics should have a small focal length, f, since $a \propto f^2$. However the reflectivity increases with decreasing graze angle, α and since $f \propto \frac{1}{\alpha}$, to achieve the highest throughput (efficiency) the optics should have the maximum possible focal length. In addition, the design has to take into account that the performance of the optics have a complex dependence on the incident photon energy and α. Moreover the reflectivity also depends on the coating material and in the number of reflections.

The fabrication technology that has been chosen is a segmented, slumped glass optics were several individual pieces of substrate make a single layer. The reasons are that the technology is mature and not expensive, it has been developed by members of IAXO for the NuStar Satellite mission [10], it is easy to deposit a single layer or multi-layers reflective coatings and the imaging requirements can be satisfied.

The design studies have taken into account the axion spectrum and the detector efficiency in order to achieve the highest figure of merit, f_{DO} with the optimisation of the optics parameters (f, a, number of layers, single and multilayer coatings). The optimal figure of merit is found for 5 m focal length with 123 layers with W/B_4C multilayers coating and with a spot of ~ 4 mm.

2.3. Low background X-ray detectors
The Micromegas detectors developed for the CAST experiment are the baseline technology for the X-ray detectors of IAXO. The levels achieved resulted from a global approach where the improvement has come from different fronts: from the manufacturing technique leading to highly intrinsically radiopure detectors with high performance, from the optimisation of the passive and

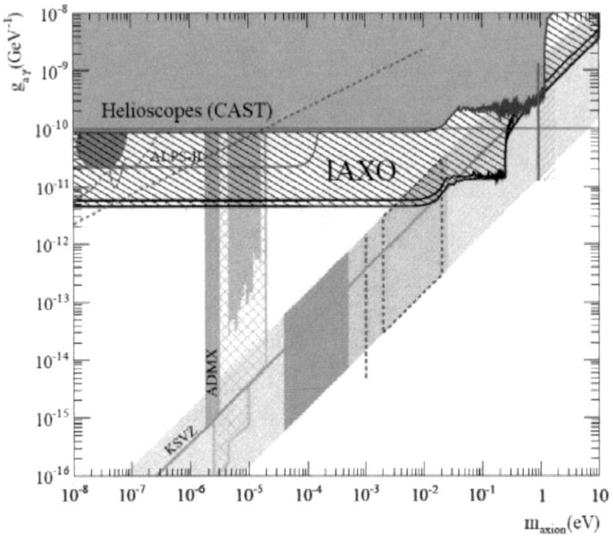

Figure 3. Expected IAXO sensitivity compared to CAST and ADMX limits. Future prospects of ADMX (dashed brown) and ALPS-II (light blue line) are shown. The nominal scenario takes into account the figures of merit described in the text while the enhanced scenario expects a lower background level (10^{-8}counts keV^{-1} cm^{-2}s^{-1} instead of 5×10^{-8}) for the Micromegas detector and an increased efficiency for the optics (70% instead of 50%) with a smaller spot size a (0.15 cm^2 instead of 0.2 cm^2).

active shielding thanks to an understanding of the background and from the refinement of the selection algorithms. The detector installed in the 2014 CAST data taking campaign in the sunrise side coupled to the X-ray telescope achieved levels 0.85×10^{-6}counts keV^{-1} cm^{-2}s^{-1} [16]. It can be considered as the first Micromegas prototype for IAXO. Further improvements are still expected:

- Use of segmented mesh microbulks[14] with auto-trigger electronics[15];
- more stringent radioactive requirements for the components of the detectors;
- upgrade of the active muon and gamma shielding;
- improvement of the selection algorithms.

3. Expected sensitivity

IAXO sensitivity, in terms of detectable signal counts, is at least 10^4 better than CAST phase-I, resulting in a sensitivity of $g_{a\gamma}$ one order of magnitude better. Simulated sensitivities for different set of parameters are given in figure 3. This exclusion plot is obtained by considering different scenarios of the expected performances of the different sub-systems of IAXO (magnet, optics, X-ray detectors). To obtain this plot a vacuum run of 3 years effective data taking, IAXO run I, is considered with sensitivities to $m_a < 0.1$ eV followed by IAXO run II that will use a buffer gas, ^{4}He, inside the magnet bores to reach higher masses up to $m_a \sim 0.25$ eV. In this way, a large unexplored region of the QCD axion phase space will be scanned reaching a few $\sim g_{a\gamma}$ 10^{-12} GeV^{-1} for masses up to 0.25 eV. At high masses, the expected sensitivity will supersede the SN 1987 energy loss limit. This region is interesting because it is favoured for DM candidates.

3.1. Additional equipment for extended sensitivity
Additional equipment can be considered to enhance the physics goals of the experiment. GridPix detectors [17], Transition Edge Sensors (TES) [18] and Charged Coupled Devices (CCD) [19] could be interesting to search for signatures requiring lower energy thresholds like hidden photons, chameleons or other ALPs. Furthermore possibilities to search for relic axions in IAXO with microwave cavities or dark matter axions with dish antennas have also been explored [11, 20]. In this way, the IAXO sensitivity could be extended to very low masses entering the region of interest for Dark Matter axions.

4. Conclusions and Status
The increasing experimental and theoretical efforts in the axion searches demonstrate that the axion search has a strong physics case. The CAST experiment has been a very important milestone in axion research during the last decade. IAXO, the fourth generation helioscope, will allow the community to deeply probe into the unexplored axion-ALP parameter space becoming a generic axion facility with discovery potential. The next short term objectives are to prepare a Technical Design Report and to construct the first prototypes of the hardware main ingredients: demonstration coil IAXO-T0, X-ray optics IAXO-X0 and low background detector IAXO-D0 while refining the physics case and studying the feasibility studies for Dark Matter axions.

Acknowledgments
We acknowledge support from the Spanish Ministry of Science and Innovation (MICINN) under contract FPA2008-03456 and FPA2011-24058, as well as under the CPAN project CSD2007-00042 from the Consolider-Ingenio2010 program of the MICINN. Part of these grants are funded by the European Regional Development Fund (ERDF/FEDER). We also acknowledge support from the European Commission under the European Research Council T-REX Starting Grant ERC-2009- StG-240054 of the IDEAS program of the 7th EU Framework Program. Part of this work was performed under the auspices of the U.S. Department of Energy by Lawrence Livermore National Laboratory under Contract DE-AC52-07NA27344 with support from the LDRD program through grant 10-SI-015. We also acknowledge support from the CAST collaboration. The design work on the magnet system was supported by CERN, Physics Department as well as the ATLAS Collabo- ration. Partial support by the Deutsche Forschungsgemeinschaft (Germany) under grant EXC-153 by the MSES of Croatia and the Russian Foundation for Basic Research (RFBR) is also acknowledged. F. I. acknowledges the support from the Eurotalents program.

References
[1] R D Pecci and Quinn H R 1977 *Phys. Lett.* **38**, 144.
[2] Sikivie P 1983 *Phys. Lett.*, 51, 1415
[3] Zioutas K *et al.* [CAST Collaboration] 2005 *Phys. Rev. Lett.* 94 121301.
[4] Andriamonje S *et al.* [CAST Collaboration] 2007 *J. Cosmol. Astropart. P.* 10 0704.
[5] Arik E *et al.* [CAST Collaboration] 2009 *J. Cosmol. Astropart. P.* 008 0902.
[6] Arik E *et al.* [CAST Collaboration] 2011 *J. Cosmol. Astropart. P.* 107 261302.
[7] Arik M *et al.* [CAST Collaboration] 2014 *J. Cosmol. Astropart. P.* 112 9, 091302.
[8] Arik M *et al.* [CAST Collaboration] 2015 arXiv:1503.00610 [hep-ex].
[9] Irastorza I G 2011 *JCAP* 1106 013.
[10] Harrison F A *et al.* 2005 *Exp. Astron.* 20 13.
[11] Irastorza I G *et al.* 2013 *CERN-SPSC-2013-022, SPSC-I-242.*
[12] Armengaud E *et al.* 2014 *JINST* 9 T05002.
[13] Aune S *et al.* 2014 *JINST* 9 01, P01001.
[14] Geralis T *et al.* 2014 *PoS TIPP2014* 055.
[15] Anvar S *et al.*, 2011 Proc. IEEE Nuclear Science Symp., 745749.
[16] Garza J G in these proceedings.

[17] Krieger C *et al.*, 2014 *PoS TIPP2014* 060.
[18] Smith S J *et al.*, 2009 *IEEE Trans. Appl. Supercond.* 19 45.
[19] Barreto J *et al.* [DAMIC collaboration], 2012, *Phys. Lett. B* 711 264 [arXiv:1105.5191].
[20] Horns D *et al.*, 2013 *J. Cosmol. Astropart. P.* 16 1304.

Cryogenic gaseous photomultipliers and liquid hole-multipliers: advances in THGEM-based sensors for future noble-liquid TPCs

Arazi L[1,4], Coimbra A E C[1,2], Erdal E[1], Israelashvili I[1,3], Rappaport M L[1], Shchemelinin S[1], Vartsky D[1], dos Santos J M F[2] and Breskin A[1]

[1] Weizmann Institute of Science, Rehovot 76100, Israel
[2] Coimbra University, Coimbra, Portugal
[3] Nuclear Research Center Negev, Beer-Sheva 9001, Israel

Email: lior.arazi@weizmann.ac.il

Abstract. Dual-phase noble-liquid TPCs are presently the most sensitive instruments for direct dark matter detection. Scaling up existing ton-scale designs to the multi-ton regime may prove to be technologically challenging. This includes both large-area coverage with affordable high-QE UV-photon detectors, and maintaining high precision in measuring the charge and light signals of rare events with keV-scale energy depositions. We present our recent advances in two complementary approaches to these problems: large-area cryogenic gaseous photomultipliers (GPM) for UV-photon detection, and liquid-hole multipliers (LHM) that provide electroluminescence light in response to ionization electrons and primary scintillation photons, using perforated electrodes immersed within the noble liquid. Results from a 10 cm diameter GPM coupled to a dual-phase liquid-xenon TPC demonstrate the feasibility of recording – for the first time – both primary ("S1") and secondary ("S2") scintillation signals, over a very broad dynamic range. The detector, comprising a triple-THGEM structure with CsI on the first element, has been operating stably at 180 K with gains larger than 10^5; it provided high single-photon detection efficiency – in the presence of massive alpha-particle induced S2 signals; S1 scintillation signals were recorded with time resolutions of 1.2 ns (RMS). Results with the LHM operated in liquid xenon yielded large photon gains, with a pulse-height resolution of 11% (RMS) for alpha-particle induced S2 signals. The detector response was stable over several months. The response of the S2 signals to rapid changes in pressure lead to the conclusion that the underlying mechanism for S2 light is electroluminescence in xenon bubbles trapped below the immersed THGEM electrode. Both studies have the potential of paving the way towards new designs of *dual-* and *single-phase* noble-liquid TPCs that could simplify the conception of future multi-ton detectors of dark matter and other rare events.

1. Introduction

Noble-liquid detectors used in dark matter searches and neutrino physics experiments are ever increasing in size and complexity. In particular, dual-phase noble-liquid time projection chambers (TPCs) [1-3], presently leading the effort for direct dark matter detection, are now entering the ton-scale regime with XENON1T [4] under construction at LNGS and ArDM [5] undergoing commissioning at LSC. Other ton-scale programs for direct dark matter detection with dual-phase

[4] Corresponding author.

liquid xenon (LXe) and liquid argon (LAr) TPCs include LZ [6], PandaX [7] and DarkSide G2 [8]. These detectors comprise a large volume of liquid target (liquid xenon, LXe or liquid argon, LAr), in equilibrium with its vapor phase. Radiation interactions within the liquid lead to a prompt scintillation signal ("S1"). Ionization electrons liberated at the site of interaction drift towards the liquid surface; they are extracted into the gas phase, inducing, through electroluminescence, a delayed scintillation signal ("S2") - proportional to the number of extracted electrons. Both S1 and S2 signals are usually recorded by top and bottom arrays of vacuum photomultiplier tubes (PMT). S1 usually defines the detection threshold; S1 and S2 are used to estimate the deposited energy and provide the 3D event localization. The ratio S2/S1 permits gamma background suppression (in LXe); in LAr detectors this is done by pulse-shape analysis.

The next step in the evolution of dual-phase TPCs will take them in the 2020s into the multi-ton regime with programs such as DARWIN [9], where they will probe the WIMP mass-cross section parameter space down to the region where irreducible neutrino background dominates. The realization of multi-ton dual-phase noble-liquid TPCs involves technological challenges that go beyond a simple scale-up of existing ton-scale designs. Among these challenges are the need to cover very large areas with affordable high-QE, low-radioactivity photon detectors (possibly in $\sim 4\pi$ configuration), and the requirement for maintaining high precision in measuring the charge and light signals of keV-scale events, over a few-meter diameter TPC. In what follows we discuss our recent advances in two complementary approaches to these challenges: large-area cryogenic gaseous photomultipliers (GPM) for UV-photon detection, and liquid hole-multipliers (LHM), first suggested in [10], in which scintillation light is produced in perforated electrodes (e.g. THGEMs or GEMs) immersed within the noble liquid.

GPMs are high-gain gaseous detectors, in which incoming photons release photoelectrons from a high-QE photocathode; these are subsequently focused into a region of a strong electric field, where they undergo avalanche multiplication, enabling high single-photon detection and localization capabilities [11]. GPMs comprising a reflective UV-sensitive cesium iodide (CsI) photocathode [12] coupled to either wire chambers [13] or cascaded gas electron multipliers (GEM [14, 15]), have been successfully employed in room temperature particle physics experiments as RICH devices [13, 15-17]. An extensive R&D has been done on CsI-coated Thick-GEM (THGEM) GPMs [18-22], in particular for the COMPASS experiment [23,24].

In the context of dark matter detection, cryogenic GPMs [25] may allow for large-area coverage with high detection efficiency and high filling factor, potentially at significantly lower cost than PMTs and with a similar level of radio-purity. In dual-phase TPCs, GPMs with reflective CsI photocathodes may serve as the top array of photon sensors. With the possibility to segment their readout to small pixels, they can allow for an improved position reconstruction capability, with potential benefits in calibration and background rejection. Furthermore, GPMs with a semi-transparent photocathode may be mounted on the TPC wall (behind the field-shaping rings) to allow for close-to 4π coverage, with a potentially large increase in the detector sensitivity to low-energy depositions, and hence low-mass WIMPs.

A primary concern in scaling up dual-phase TPCs to the multi-ton regime is the requirement to keep the liquid-gas interface completely parallel to the two meshes bounding it from above and below across several meters. Especially for LXe TPCs, this requirement is essential for maintaining a high level of background discrimination capability. A possible solution to this problem may be to devise a *single-phase* liquid-only TPC, in which the charge signal (S2) is produced within the liquid by immersed electrodes rather than in the vapor phase. While other groups have investigated the feasibility of this idea by using thin wires [26, 27], we have chosen to focus on immersed liquid hole-multipliers (LHM) [10] (e.g. THGEMs or GEMs). Preliminary

Journal of Physics: Conference Series **650** (2015) 012010 doi:10.1088/1742-6596/650/1/012010

results with a THGEM immersed in LXe were highly encouraging, with surprisingly large light gains, proportional to the THGEM field [28].

In this work we summarize our recent results in operating a 10 cm diameter, triple-THGEM cryogenic GPM prototype with a reflective CsI photocathode, coupled through a UV window to a small dual-phase LXe TPC. These results represent the first demonstration of the ability of such a detector to record both single photons and massive S2 signals in the same operating conditions. We also summarize our recent advances with the LHM concept, including a surprising new understanding of its underlying electroluminescence mechanism. Detailed reports about both studies can be found in [29] and [30].

2. Cryogenic gaseous photomultipliers

The experiments were conducted using the Weizmann Institute Liquid Xenon (WILiX) setup, described in detail in [27]. The cryostat comprises an inner vacuum-insulated LXe-filled chamber which housed, for the GPM experiment, a small dual-phase TPC at its center. Throughout the GPM experiments the cold finger condensing the xenon vapor was kept at 170 K, with the xenon gas pressure fixed at 1.8 bar; xenon gas was continuously recirculated at 3 slpm through an SAES hot getter for purification.

The TPC is schematically shown in figure 1 (a). It is comprised of two electro-formed Cu meshes set 5 mm apart, serving as the anode and gate and bounding the liquid-gas interface from above and below, respectively. A stainless steel disc, serving simultaneously as the TPC cathode and a spectroscopic alpha particle source (^{241}Am with an activity of 80 Bq), was held 5 mm below the gate mesh. The TPC voltages were set to create a drift field of 1 kV/cm between the source and gate and ~10 kV/cm between the gate and anode. Alpha emissions from the source into the liquid resulted in prompt S1 light signals, followed by secondary S2 signals appearing ~2.4 μs later. A 1" square PMT (Hamamatsu R8520-06-Al), located 3.5 cm below the top surface of the source, was used to record reflected S1 and S2 photons.

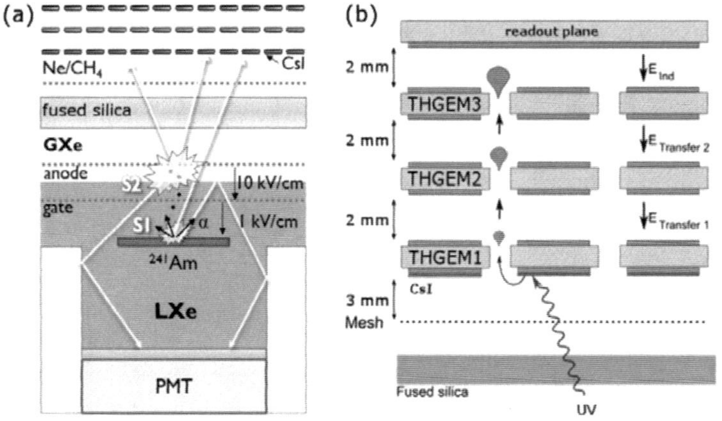

Figure 1: (a) Schematic drawing of the experiment, showing alpha particle-induced S1 and S2 signals detected directly by the GPM and, through reflections, by the bottom PMT. (b) The triple-THGEM GPM scheme.

The GPM assembly (figure 1 b) was installed above a fused silica window. It comprised a cascaded structure of three THGEMs held 2 mm apart, with an active diameter of 100 mm. Each consisted of a 0.4 mm thick FR4 plate, Cu-clad on both sides with an hexagonal pattern of 0.4 mm diameter holes surrounded by 50 μm etched rims, with 0.8 mm between their centers. CsI deposition on the first THGEM was carried out in a dedicated evaporation setup, following

Journal of Physics: Conference Series **650** (2015) 012010 doi:10.1088/1742-6596/650/1/012010

standard procedures [12]. The deposited layer was ~300 nm thick with a nominal QE of ~20-25% at 175 nm. The signals were recorded from the whole, un-segmented readout plane, with a charge-sensitive preamplifier located outside the system. An electro-formed Cu mesh with 85% transparency was set 3 mm below the first THGEM and kept at the same potential as its CsI-coated face (to optimize the extraction of photoelectrons from the photocathode). The GPM working gas was Ne/CH_4, chosen because it provides both a high gas gain at relatively low voltages [31] and high photoelectron extraction efficiencies [20, 32]. For the initial experiments we used Ne/CH_4(5%) and Ne/CH_4(20%). Most of the data were taken at a pressure of 0.7 bar with the GPM gas at 180 K. Typical S1 and S2 signals of both the GPM and the reference PMT are shown in figure 2.

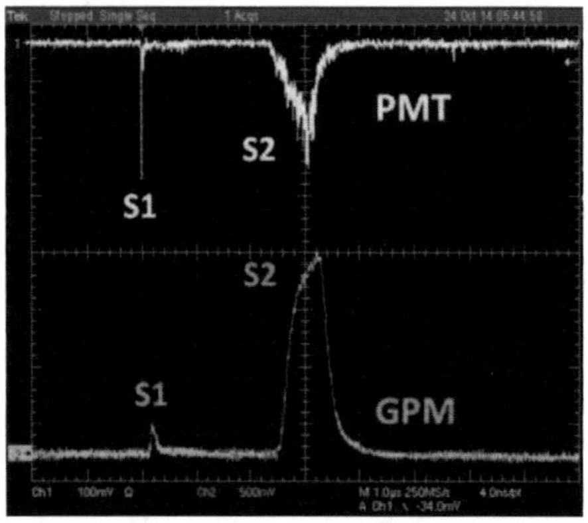

Figure 2: Typical alpha-particle induced S1 and S2 signals recorded by the PMT and GPM (shown in figure 1b). The GPM, in this case, was operated with Ne/CH_4(5%) at 0.7 bar and 180 K, at a gain of ~1×10^5. The GPM signal was shaped using a timing filter amplifier.

Figure 3 shows the measured GPM gain curves for Ne/CH_4(5%) and Ne/CH_4(20%). The voltage division was asymmetric in both cases, with the highest voltage across THGEM1 to maximize the photoelectron extraction efficiency [20] and improve the overall stability. For both gas mixtures the maximal gain was well above 10^5, assuring high single-photon detection efficiency. At a gain of 10^5 in Ne/CH_4(5%) the operation with alpha-particles at 40 Hz, yielding very large S2 pulses, was fully stable - indicating a discharge probability below 3×10^{-6}. To the best of our knowledge, this is the first time a cryogenic GPM was shown to record S1 and S2 signals from a LXe dual-phase TPC and operate stably over such a broad dynamic range - from one to several thousand photoelectrons.

Journal of Physics: Conference Series **650** (2015) 012010 doi:10.1088/1742-6596/650/1/012010

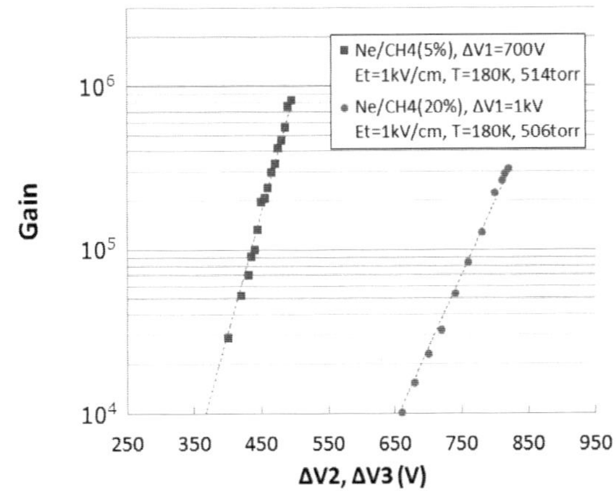

Figure 3: Gain curves of the GPM shown in figure 1b, with Ne/CH_4(5%) and Ne/CH_4 (20%) at 180 K and 0.7 bar, vs. the equal voltages across the second and third stages. The voltage values across the first THGEM, ΔV_1, are given in the inset; the induction and transfer fields were 1 kV/cm in all cases.

S1 and S2 spectra are shown in figure 4. The estimated number of photoelectrons in the GPM is a few tens for S1 and a few thousands for S2 signals. For S1 the RMS resolution is σ/E ~11% (by fitting a Gaussian to the entire S1 peak). For S2, the asymmetric shape of the spectrum reflects the coincidence of alpha and 59.5 keV gamma emissions from the source, and a Gaussian fit to the left side of the peak gives σ/E~8.7%. The S2 resolution recorded here is roughly the same as that obtained in XENON100 (σ/E=10.0±1.5%) with PMTs, for a similar number of ionization electrons entering the gas phase [33]. The ratio of alpha and 59.5 keV peaks (~4) is consistent with the vastly different charge yields of ionization electrons from the tracks of an alpha particle and of a 60 keV electron [3].

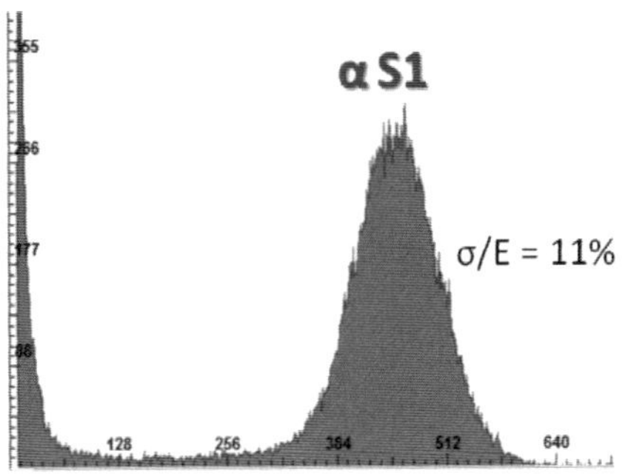

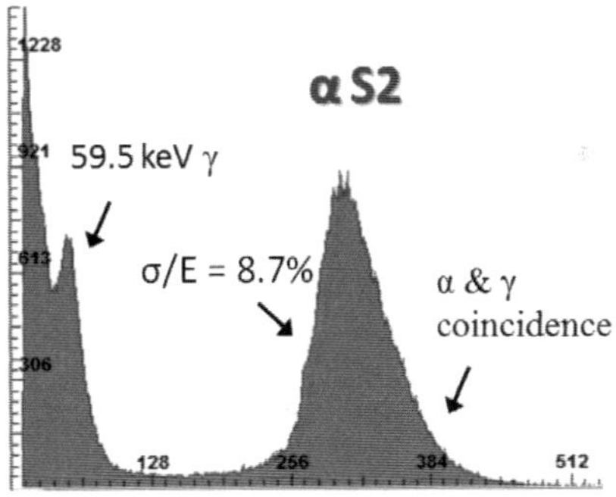

Figure 4: Pulse-height spectra recorded with the GPM in the setup shown in figure 1, with a spectroscopic ^{241}Am source. Left –S1 spectrum; right –S2 spectrum. Note the 59.5 keV gamma peak and the asymmetric right side of the alpha-particle S2 peak, attributed to coincidence between alpha and gamma emissions. Data recorded with Ne/CH_4(5%) at 180 K and 0.7 bar with a GPM gain of 1×10^5.

Figure 5 (left) shows the PMT alpha-particle S1 signal and the corresponding GPM one (after the charge-sensitive preamplifier), with the trigger on the PMT signal. The ~200 ns delay in the GPM signal reflects the time lapse from photoelectron extraction to the collection of the avalanche electrons on the readout plane. The distribution of the time difference between the two pulses (figure 5, right), defines the time resolution of the GPM (here – for a few dozen photoelectrons); it is ~1.2 ns.

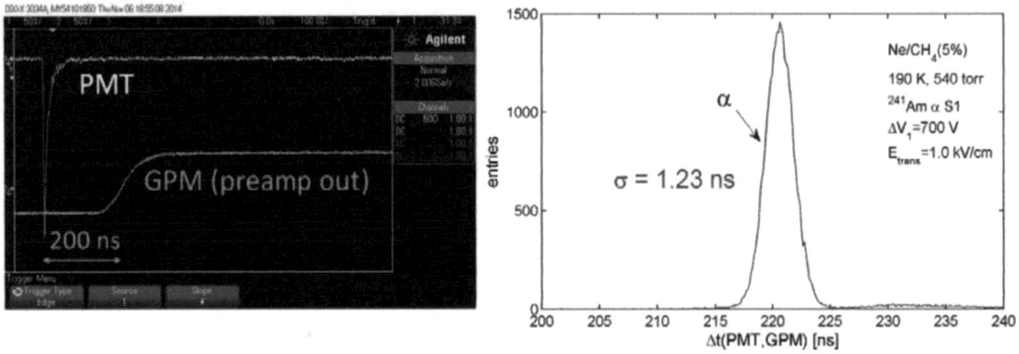

Figure 5: Left: alpha-particle S1 signals recorded by the PMT and GPM. Right: PMT-GPM time-difference distribution. The RMS width, defining the GPM time resolution, is ~1.2 ns.

3. Liquid hole-multipliers

The preliminary results for alpha-induced secondary scintillation ("S2") in a THGEM immersed in LXe were reported in [28]; they indicated a large light gain of a few hundred photons per ionization electron focused into the THGEM holes. However, the very low threshold field in the holes (a few kV/cm) required for the appearance of S2 signals led us to suspect that the scintillation mechanism was different than originally thought. Extensive investigations, described in detail in [30], confirmed the initial observations of steady-state S2 signals, but we now have the following understanding: *the observed electroluminescence occurs in gas bubbles trapped below the THGEM electrode, rather than in the liquid itself.* A schematic drawing of the experiment (installed in WILiX), in light of this new understanding, is shown in figure 6.

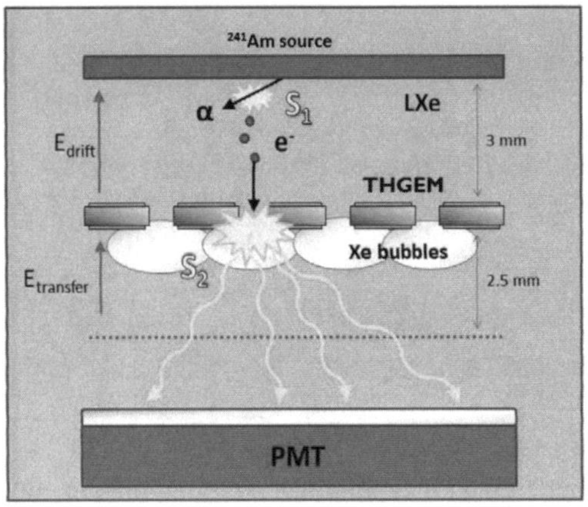

Figure 6: Schematic drawing of the LHM setup, including the (hypothesized) xenon gas bubbles trapped below the THGEM. Alpha-induced ionization electrons drift towards the THGEM electrode and are focused into its holes. The high field inside the holes drives them into the trapped gas bubbles underneath, where they produce the S2 light, through electroluminescence - recorded by the PMT.

Aside from the very low threshold field for scintillation (smaller by about two orders of magnitude compared to the values reported for thin wires immersed in LXe [26, 34, 35]), the most compelling evidence for bubbles is the response of the S2 signals to rapid changes in the system pressure. A rapid increase in pressure (induced by a step increase in the cold finger temperature), results in the immediate disappearance of the S2 signals; the signals reappear either if the pressure is reduced back to its original value, or if the system is allowed to warm up until stabilizing at a higher steady-state temperature. Our interpretation is as follows. When the system is in thermodynamic steady state, bubbles are continuously formed on one or more 'hot' surfaces, and are subsequently trapped below the THGEM - allowing for the generation of electroluminescence signals in the sufficiently high THGEM field. A rapid pressure increase causes the bubbles to collapse; bubble formation stops because the temperature of the bubble-forming surfaces (which does not change sufficiently fast) is too low to allow bubbles to grow against the new, higher pressure. Bubble formation resumes when the temperature of these surfaces becomes high enough, resulting in the reappearance of S2.

During several months of experiments, S2 signals were always present when the system was in thermodynamic steady state (i.e., fixed temperature and pressure), appearing for ~99.5% of the triggers on alpha-particle induced S1 - indicating a constant presence of bubbles below the THGEM electrode. The magnitude of S2 signals, shown in figure 7a, depended approximately linearly on the THGEM voltage; it remained practically unchanged during 3 months of measurements; surprisingly, no pressure-dependence was observed. The RMS width (σ/E) of the alpha-particle S2 in steady state conditions was ~20-25% as a result of occasional S2 amplitude fluctuations. However, when the S2 signals reappeared (after gradual warming up of the system following an increase in pressure), they remained, for 1-2 hours, in a 'super-stable' state, in which their RMS resolution was ~11%, close to that of XENON100 for the same number of drifting electrons (σ/E=10.0±1.5%) [33]. The spectrum of S2 signals in the 'super-stable' state is shown in figure 7b.

Assuming that the major contribution to S2 comes from light emitted at the bottom of the THGEM holes, we estimate that the light gain is a few tens of photons emitted into 4π per electron focused into the hole, for a THGEM voltage of 3 kV.

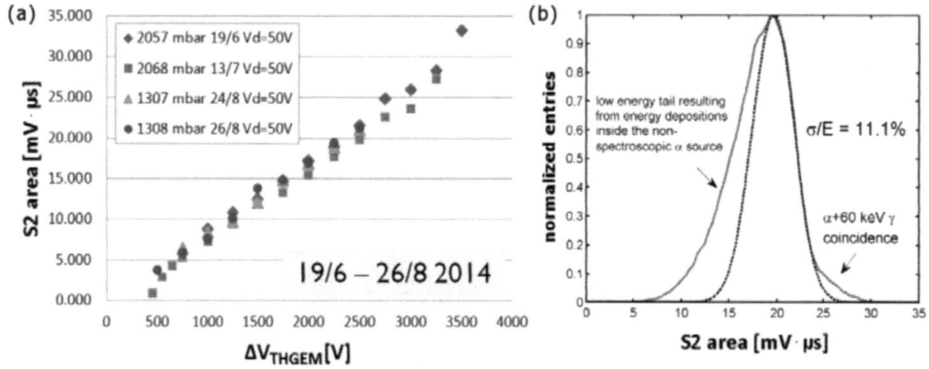

Figure 7: (a) S2 area (time integral of pulse) vs. the THGEM voltage, for a drift voltage of 50 V (corresponding to a drift field of ~0.2 kV/cm), taken at different dates over more than two months; note the surprising similarity between the results at 1.3 bar and 2.1 bar. (b) Alpha-particle S2 area distribution, right-fitted with a Gaussian. The low-energy tail on the left reflects the fact that the ^{241}Am source used in these measurements was non-spectroscopic (unlike the one used in the GPM study). The RMS resolution is 11.1%, close to that of XENON100 for the same number of ionization electrons.

4. Summary and discussion

The present study provides a first proof-of-concept for using cryogenic GPMs as large-area detectors, capable of recording both S1 and S2 signals from a LXe dual-phase TPC over a broad dynamic range. The demonstrated S2 energy resolution recorded by the GPM for alpha particles is similar to that obtained with XENON100 using PMTs for the same number of ionization electrons. The time resolution (here measured for pulses comprising a few dozen photoelectrons) is ~1.2 ns, well within the requirements of direct dark-matter detection experiments. While the GPM single-photon detection efficiency was not measured directly in this experiment, based on the QE of CsI at 175 nm (25%), the geometry of the electrodes, the mesh transparency and the gas composition, we estimate it to be ~10%, in the present setup. This can be further increased to ~15% by optimizing the detector parameters as discussed in [29]. Assuming a 90-95% filling factor of square GPM modules of a typical size of 30×30 cm^2, the overall single-photon detection efficiency for an entire GPM array would be ~14%. We note that the GPM single-photon detection efficiency may be further enhanced by the addition of a semi-transparent CsI photocathode on the inner surface of the fused-silica window. This option, also important for 4π coverage of the TPC (figure 8a), is currently under investigation by our group. The GPM concept has been also investigated by us as a readout element for a LXe camera for combined fast-neutron and gamma imaging [36, 37]

Our recent advances in understanding the underlying mechanism for electroluminescence in a THGEM immersed in LXe, and in particular the identification of the 'super-stable' boiling regime with its accompanying high S2 RMS resolution, may open new possibilities for the design of future noble-liquid TPCs. An essential requirement for the successful implementation of this idea (the '*bubble-assisted LHM*') is to devise a scheme in which bubbles are produced in a controllable manner. A successful outcome could lead to the conception of single-phase liquid-only TPC designs, in which ionization electrons drift down towards the bubble-assisted LHM (figure 8b). Photons could be recorded with pixilated GPM arrays underneath. Furthermore, coating the LHM top surface with CsI would allow for the detection of S1 photons as well as ionization electrons, as originally suggested in [10].

Although the present study focused on LXe, both the GPM and LHM concepts may be also considered for LAr TPCs. GPMs for LAr may be conceived with MgF_2 windows for the direct detection of 128 nm photons (without wavelength shifters), relying on the high QE of CsI at this wavelength (>55% [12]). We note that the bubble-assisted electroluminescence identified in LXe may have also been responsible for previous observations in LAr with immersed THGEM [38] and GEM [39] electrodes. Besides dark matter searches, one may thus consider the use of bubble-assisted LHMs with GPM readout in LAr TPCs for neutrino oscillation experiments. More details can be found in [29, 30].

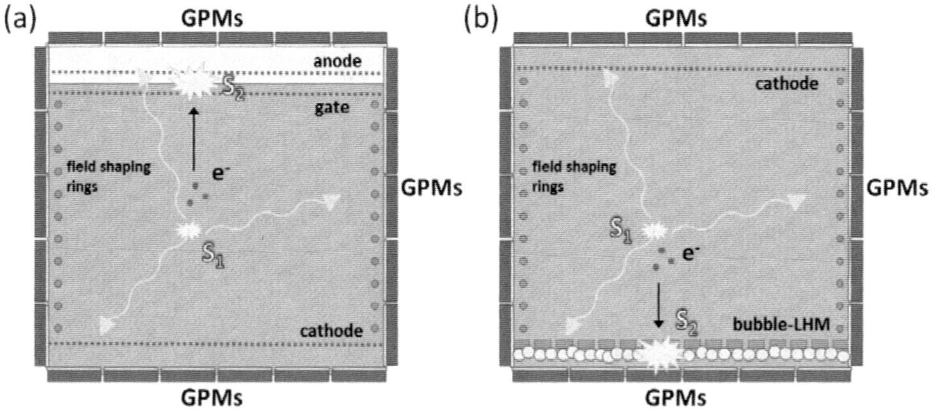

Figure 8: Possible TPC configurations employing GPMs and LHMs. (a) Dual-phase TPC with 4π GPM readout; the field shaping rings are spaced to allow photons to reach the wall GPMs. (b) Single-phase liquid-only TPC with inverted drift field, bubble-assisted LHM at its bottom and 4π GPM coverage.

Acknowledgements

This work was partly supported by the Minerva Foundation with funding from the German Ministry for Education and Research (Grant No. 710827), the Israel Science Foundation (Grant No. 477/10), the PAZY Foundation (Grant No. 258/14) and by FCT and FEDER under the COMPETE program, through project CERN/FP/123614/2011. A. Breskin is the W.P. Reuther Professor of Research in The Peaceful Use of Atomic Energy.

References

[1] Dolgoshein B A, Lebedenko V N and Rodionov B U 1970 New Method of Registration of Ionizing-particle Tracks in Condensed Matter *JETP Letters* **11**(11).
[2] Aprile E and Doke T 2010 Liquid xenon detectors for particle physics and astrophysics *Rev Mod Phys* **82**(3) 2053-97.
[3] Chepel V and Araujo H 2013 Liquid noble gas detectors for low energy particle physics 2013 *JINST* **8** R04001.
[4] Aprile E for the XENON1T collaboration 2012 The XENON1T Dark Matter Search Experiment arXiv:1206.6288 .
[5] Badertscher A, et al. 2013 ArDM: first results from underground commissioning 2013 *JINST* **8** C09005.
[6] Malling D C, et al 2011 After LUX: The LZ Program, arXiv:1110.0103
[7] Cao X G, et al. 2014 PandaX: a liquid xenon dark matter experiment at CJPL *Sci China Phys Mech* **57**(8) 1476-94.
[8] Alexander T, et al. 2013 DarkSide search for dark matter 2013 *JINST* **8** C11021.
[9] Baudis L for the DARWIN Consortium 2012 DARWIN dark matter WIMP search with noble liquids *J Phys Conf Ser* **375** 2028.
[10] Breskin A 2013 Liquid Hole-Multipliers: A potential concept for large single-phase noble-liquid TPCs of rare events *J Phys Conf Ser* **460**.

[11] Chechik R and Breskin A 2008 Advances in gaseous photomultipliers *Nucl Instrum Meth A* **595**(1) 116-27.
[12] Breskin A 1996 CsI UV photocathodes: History and mystery *Nucl Instrum Meth A* **371**(1-2) 116-36.
[13] Piuz F 2003 Ring Imaging CHerenkov systems based on gaseous photo-detectors: trends and limits around particle accelerators *Nucl Instrum Meth A* **502**(1) 76-90.
[14] Sauli F 1997 GEM: A new concept for electron amplification in gas detectors *Nucl Instrum Meth A* **386**(2-3) 531-4.
[15] Mormann D, Breskin A, Chechik R, Cwetanski P and Singh B K 2002 A gas avalanche photomultiplier with a CsI-coated GEM *Nucl Instrum Meth A* **478**(1-2) 230-4.
[16] Kozlov A, Ravinovich I, Shekhtman L, Fraenkel Z, Inuzuka M and Tserruya I 2004 Development of a triple GEM UV-photon detector operated in pure CF4 for the PHENIX experiment *Nucl Instrum Meth A* **523**(3) 345-54.
[17] Fraenkel Z, et al. 2005 A hadron blind detector for the PHENIX experiment at RHIC *Nucl Instrum Meth A* **546**(3) 466-80.
[18] Chechik R, Breskin A, Shalem C and Mormann D 2004 Thick GEM-like hole multipliers: properties and possible applications *Nucl Instrum Meth A* **535**(1-2) 303-8.
[19] Breskin A, et al. 2009 A concise review on THGEM detectors *Nucl Instrum Meth A* **598**(1) 107-11.
[20] Azevedo C D R, et al. 2010 Towards THGEM UV-photon detectors for RICH: on single-photon detection efficiency in Ne/CH4 and Ne/CF4 2010 *JINST* **5** P01002.
[21] Peskov V, Cortesi M, Chechik R and Breskin A 2010 Further evaluation of a THGEM UV-photon detector for RICH – comparison with MWPC 2010 *JINST* **5** P11004.
[22] Breskin A, et al. 2011 CsI-THGEM gaseous photomultipliers for RICH and noble-liquid detectors *Nucl Instrum Meth A* **639**(1) 117-20.
[23] Alexeev M, et al. 2012 Development of THGEM-based Photon Detectors for COMPASS RICH-1 *Physcs Proc* **37** 781-8.
[24] Alexeev M, et al. 2013 Status and progress of novel photon detectors based on THGEM and hybrid MPGD architectures 2013 *JINST* **8** C12005.
[25] Duval S, et al. 2012 Hybrid multi micropattern gaseous photomultiplier for detection of liquid-xenon scintillation *Nucl Instrum Meth A* **695** 163-7.
[26] Aprile E, et al. 2014 Measurements of proportional scintillation and electron multiplication in liquid xenon using thin wires 2014 *JINST* **9** P11012.
[27] Ye T, Giboni K L and Ji X 2014 Initial evaluation of proportional scintillation in liquid Xenon for direct dark matter detection 2014 *JINST* **9** P12007.
[28] Arazi L, et al. 2013 First observation of liquid-xenon proportional electroluminescence in THGEM holes 2013 *JINST* **8** C12004.
[29] Arazi L, et al. 2015 First results of a large-area cryogenic gaseous photomultiplier coupled to a dual-phase liquid xenon time projection chamber, submitted to *JINST*.
[30] Arazi L, et al. 2015 Bubble-assisted electroluminescence in the holes of a THGEM immersed in liquid xenon, submitted to *JINST*.
[31] Cortesi M, et al. 2009 THGEM operation in Ne and Ne/CH4 2009 *JINST* **4** P08001.
[32] Escada J, Coelho L C C, Dias T H V T, Lopes J A M, Santos J M F d and Breskin A 2009 Measurements of photoelectron extraction efficiency from CsI into mixtures of Ne with CH4 and CF4 2009 *JINST* **4** P11025.
[33] Aprile E, et al. 2012 The XENON100 dark matter experiment *Astropart Phys* **35**(9) 573-90.

[34] Lansiart A, Seigneur A, Moretti J L and Morucci J P 1976 Development Research on a Highly Luminous Condensed Xenon Scintillator *Nucl Instrum Methods* **135**(1) 47-52.

[35] Masuda K, et al. 1979 Liquid Xenon Proportional Scintillation-Counter *Nucl Instrum Methods* **160**(2) 247-53.

[36] Breskin A, et al. 2012 A novel liquid-Xenon detector concept for combined fast-neutrons and gamma imaging and spectroscopy 2012 *JINST* **7** C06008.

[37] Israelashvili I, et al. 2015 A Comprehensive Simulation Study of a Liquid-Xe Detector for Contraband Detection, 2015 *JINST* **10** P03030.

[38] Lightfoot P K, Barker G J, Mavrokoridis K, Ramachers Y A and Spooner N J C 2009 Optical readout tracking detector concept using secondary scintillation from liquid argon generated by a thick gas electron multiplier 2009 *JINST* **4** P04002.

[39] Buzulutskov A 2012 Advances in Cryogenic Avalanche Detectors 2012 *JINST* **7** C02025.

Recent R&D results on LAr LEM TPC and plans for LBNO demonstrators

C. Cantini, L. Epprecht, A. Gendotti, S. Horikawa, S. Murphy, G. Natterer, L. Periale, C. Regenfus, F. Resnati, A. Rubbia, F. Sergiampietri, T. Viant and S. Wu
on behalf of the LAGUNA-LBNO and WA105 collaboration

ETH Zurich, Institute for Particle Physics, CH-8093 Zürich, Switzerland

E-mail: Shuoxing.Wu@cern.ch

Abstract. The double phase Liquid Argon (LAr) Time Projection Chamber (TPC) is the state-of-art technology for neutrino detection thanks to its superb 3 Dimensional (3D) tracking and calorimetry performance. Based on this technology, the Giant Liquid Argon Charge Imaging ExpeRiment (GLACIER) is proposed to be the far detector for the Long Baseline Neutrino Oscillation (LBNO) experiment aiming at studying neutrinos 2300 km away from their production point. We report recent R&D results on the charge readout system for GLACIER and the plans to build the GLACIER demonstrators at CERN.

1. Introduction

With the last mixing angle θ_{13} measured [1, 2, 3], neutrino physics enters a new era. Thanks to the large θ_{13} value, determining the ordering of the 3 neutrino mass eigenstates (mass hierarchy) and measuring the charge parity violation (CPV) phase δ_{CP} of the Pontecorvo-Maki-Nakagawa-Sakata (PMNS) matrix become possible. The proposed LBNO [4] experiment will perform these measurements by studying the $\nu_\mu \to \nu_e$ and $\bar{\nu}_\mu \to \bar{\nu}_e$ oscillations. The LBNO experiment fully exploits the dependence of the oscillation probability as a function of L/E. The energy spectrum of the neutrino beam and the 2300 km baseline are chosen so that the $\nu_\mu \to \nu_e$ oscillation covers its first and second maxima. With its first 5 years running with GLACIER 20 kton as the far detector, LBNO could determine the mass hierarchy to 5σ confidence level (C.L.) over the full δ_{CP} range and provide evidence of CPV at 3σ C.L. [5]. With 10 years running with GLACIER 20 and 50 kton as the far detectors and a high power PS beam, LBNO could determine δ_{CP} at 5σ C.L. over 54% of the true δ_{CP} value [5]. Besides studying accelerator neutrinos, LBNO also has the unprecedented potential for detection of atmospherical and solar neutrino as well as proton decay searches.

The LAr TPC has the unique advantage of being both a 3D tracking detector and a calorimeter. The electron and muon created from ν_e and ν_μ interaction with argon nucleus have distinct behaviours when travelling in the LAr and are easily identified. Thanks to its fine granularity, the LAr TPC has a good separation power for electron from π^0 which is one of the largest backgrounds for ν_e appearance signal. The sub-mm transverse diffusion for drifting electrons over several meters enables building giant LAr TPCs without charge imaging distortion. The 128 nm scintillation light provides a very precise event starting time for 3D track reconstruction.

Unlike traditional single phase LAr TPC using wire planes to readout the drifting electrons, GLACIER uses the micro-patten gas detector technology [6] to provide charge amplification in the gas phase. The amplified charge is then collected on a novel 2 dimensional (view) anode. Unlike the bi-polar signals from the induction views of the single phase TPC, the two views of GLACIER both have the same unipolar signals which greatly facilitates the event reconstruction. Thanks to its adjustable gain, GLACIER lowers the detection threshold and covers a broad energy range from MeV to multi-GeV. This feature enables GLACIER to be also sensitive to solar and super-nova neutrinos. Despite the many appealing advantages, the technical challenges towards GLACIER need intensive R&D and prototyping efforts.

2. R&D results of the charge readout system

The charge readout system of GLACIER 20 and 50 kton detector is in a modularised configuration [7]: each modular is a 4×4 m^2 charge readout plane (CRP) consisting of 64 50×50 cm^2 anodes and Large Electron Multiplier (LEM) panels respectively and one 4×4 m^2 extraction grid. 4 anodes are bridged together to form a readout length of 2 m, and each LEM is functioning independently. Issues associate with the increased detector size appear and need to be addressed: 1. the increased inter-strip capacitance will bring significant noise to the front-end electronics; 2. the increased probability of badly drilled holes of the LEM will limit the maximal gain and might prevent the LEM from stable operation due to more frequent discharges; 3. the worsen matching between the extraction grid and the anode strips will degrade the charge collection uniformity. All these issues closely relate to the charge imaging performance of GLACIER thus need to be solved on smaller prototypes.

The R&D for the GLACIER charge readout system was carried out with the so called "3 liter" setup [8]. It is a double phase LAr LEM TPC with a readout area of 10×10 cm^2 and a drift length of 21 cm. It's a small setup and could be easily operated with a fast turn-around thus very useful to test new ideas. After being evacuated for several days, the chamber is filled with filtered liquid argon and then exposed to cosmic muons for several days data taking. The high quality muon samples with a known energy lose per unit length (dE/dx) are used to calibrate the chamber in terms of effective gain, charge sharing and long term stability. The recorded data was reconstructed using the software package QScan [9] which does hit finding, clustering, 2D and 3D tracking, etc.

2.1. The low capacitance anode

As reported in [8], we successfully developed the 2D readout anode consisting of two sets of orthogonal copper strips separated by a layer of Kapton (polyamide) with a typical thickness of 50 μm. Though well functioning at a readout length of ~10 cm, this technology is not suitable to scale up to meter level as required by the GLACIER. The difficulties come from: 1. due to the small gap thus a big capacitance (~600 pF/m [10]) between the X and Y strips, the noise level of the pre-amplifier will increase dramatically; 2. since the X strips lay above the Y strips, the anode can not be produced from a standard PCB production procedure and etching the Kapton will bring extra complexity.

Figure 1 shows the novel anode layout and the schematic as the solution. The red pads and tracks denote the readout in one direction and the unfilled ones for the other direction. For both views, the pads and tracks are printed on the same layer and the connection is done through vias and tracks on the other layer. Due to the fully symmetric layout between the two views, the charge sharing asymmetry is kept within 1% as seen from Fig. 2 right. Thanks to the additional pads in between the readout pads and tracks, we are able to keep a uniform charge collection between neighbouring strips within the same view. A direct result is the similar measured $\Delta Q/\Delta s$ for tracks with different azimuth angles as shown in Fig. 2 left and middle.

The capacitance of this novel anode is measured to be ~150 pF/m, which is a factor of 4 reduction compared with the Kapton type anode. Thanks to its integration of two views on the same layer, the anode could be produced by PCB industry using standard copper etching technique. As reported in [10], we also tried other geometry layout and the one showed in this paper gave the best results in terms of capacitance and resolution.

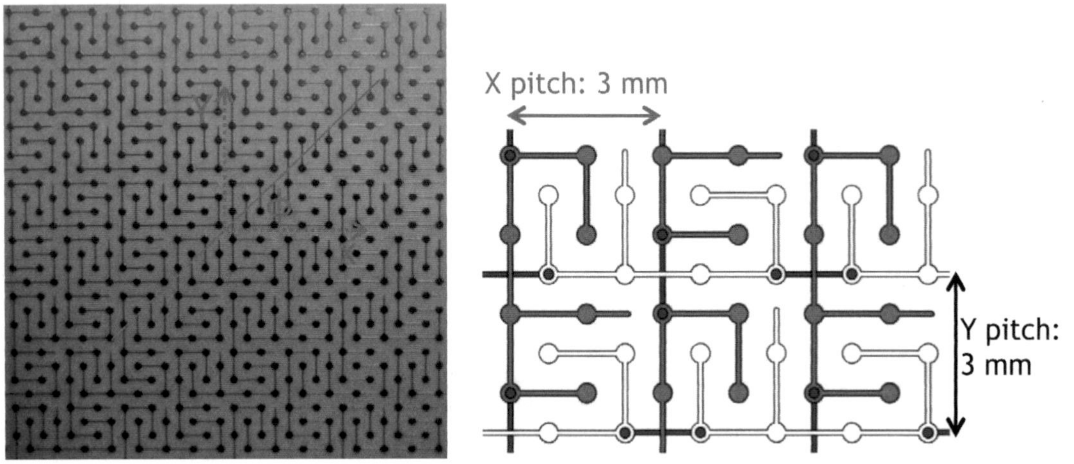

Figure 1. The novel 2D anode (left) and the schematic (right). The red pads and tracks denote one view and the white for the other view.

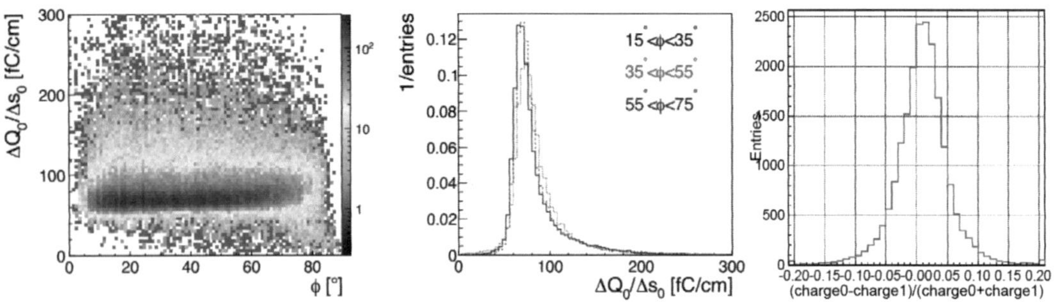

Figure 2. Performance of the novel anode. Left: the measured $\Delta Q_0/\Delta s_0$ vs. track azimuth angle; middle: distributions of $\Delta Q_0/\Delta s_0$ for three azimuth angle ranges; right: the distribution of the difference of charge collection between two views.

2.2. The LEM

The LEM or Thick GEM (THGEM) is the key apparatus of the double phase LAr TPC. Its function is to amplify the electrons produced in the LAr medium in the gas phase before being collected on the anode. Holes at a constant pitch are mechanically drilled through a copper-cladded FR4, and rims are later etched around the holes to prevent discharges occuring between the top and bottom electrodes. A high electrical potential over 3 kV is applied across the two electrodes resulting in an electric field high enough to trigger the Townsend avalanche for the entering electrons. The holes absorb the UV photons produced in the avalanche process and thus act as the mechanical quencher to prevent the photon feedback phenomenon. This property makes the LEM suitable to operate in ultra pure argon gas in the absence of quenching gas.

The design parameters of the LEM include the FR4 thickness, the hole diameter, the arrangement of the holes, and the size of the rim. These parameters affect both the gain at a certain potential difference and the maximal achievable gain. The effective gain is defined as the sum of $\Delta Q/\Delta s$ of both views divided by that of the muon. The effective gain takes account of the transparencies of the extraction grid-LEM, LEM itself, LEM-anode and the gain of the LEM. Thus, it could be described by $G_{eff}(E,p,T) = \mathcal{T} \times e^{x \cdot \alpha(p,T,E)}$ with first term the transparency, second term the gain of the LEM. x denotes the amplification length, α is the first Townsend coefficient with the form of $\alpha(p,T,E) = \frac{Ap}{T}e^{-\frac{Bp}{E}}$, where A and B are parameters depending on the gas. Fits to the gain curves as shown in Fig. 3 give trend both in the transparency and amplification length consistent with expectation from electrostatic simulations [11]. In terms of maximal achievable gain, the one with 1mm thick FR4, 500 μm hole in diameter, hexagonal hole layout and 40 μm rim gives the best result of a gain over 160.

As described in [10], we observed the behaviour of the initial decreasing and final stabilisation of the gain which is concluded to be the charging-up of the FR4 dielectric surface. In this study, we found both the charging-up time and the ratio between the final and initial gain depend on the geometry parameters. For a detailed discussion, please refer to [11]. Though charged up, the LEMs could still be operated at a stable gain over 20, which fully meets the requirement for GLACIER. One notable observation is that the LEMs have very few or no discharges over an operation of several days. This "discharge-free" behaviour promises us to step forward to the 50×50 cm^2 LEM as proposed for the GLACIER.

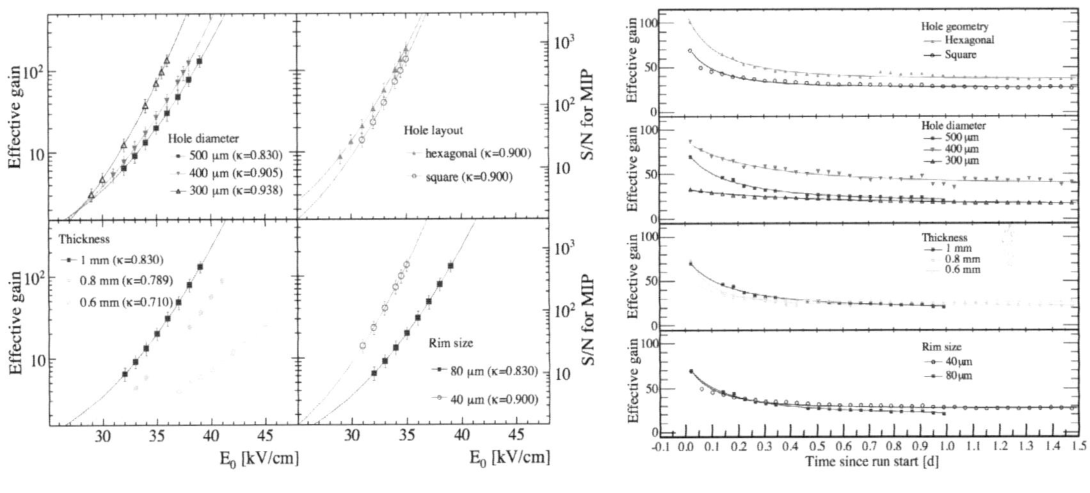

Figure 3. Performance of the LEMs with different geometry parameters. Left: effective gain vs. LEM electric field; right: the stabilisations of the effective gain over time.

2.3. The extraction grid

In the initial stage of the "3 liter" setup, there were two extraction grids placed in both liquid and gas phase [8]. Though being able to keep the charge image distortion minimised, it requires more high voltage channels and additional alignment between the two grids. As reported in [10], we successfully extracted electrons and kept the charge image undistorted by means of a single extraction grid placed in the liquid phase. By precisely aligning the stainless steel wires with

Journal of Physics: Conference Series **650** (2015) 012011 doi:10.1088/1742-6596/650/1/012011

respect to the anode strips, we are able to maintain a good gain uniformity to the level of $\pm 10\%$ as seen from Fig. 4.

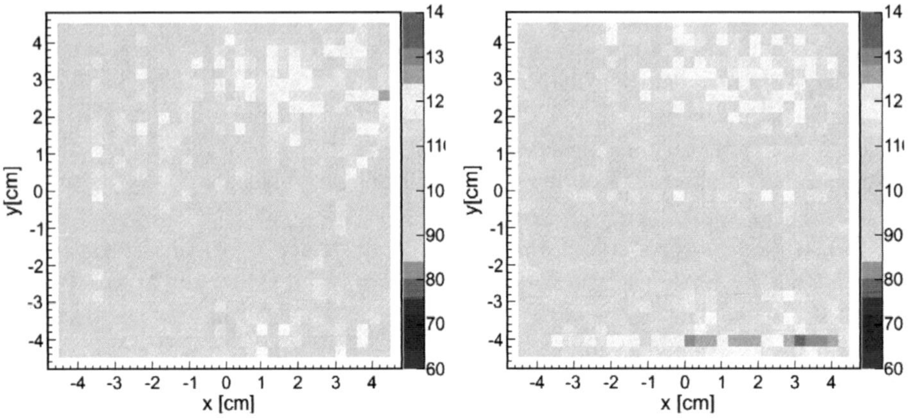

Figure 4. $\Delta Q/\Delta S$ normalised to 100 fQ/cm as a function of X and Y for view0 (left) and view1 (right).

3. Plan for the GLACIER demonstrators

With good performance achieved by the small prototypes, it's of fundamental importance to demonstrate that we could build and operate larger LAr LEM TPCs before going to the GLACIER scale. Thus, two demonstrators are proposed to demonstrate the feasibilities of the LAr LEM TPC technology in the multi-ton and multi-hundred-ton scales respectively.

3.1. The 3×1×1 m^3 demonstrator

The $3 \times 1 \times 1$ m^3 demonstrator has a fiducial volume of $3 \times 1 \times 1$ m^3 corresponding to a 4.2 ton fiducial mass. The charge readout system covers an area of 3×1 m^2 and the drift length is 1 m. The TPC is hanging from the so called "top cap" and contained inside a membrane tank with a dimension of $7.3 \times 4.9 \times 4.7$ m^3 as shown in Fig.5. The total liquid argon mass will be 20 ton. The $3 \times 1 \times 1$ m^3 demonstrator is being built and will be tested with cosmic muons at building 182 of CERN beginning of 2016. As a purely technical demonstrator, the $3 \times 1 \times 1$ m^3 will give vital feedback on the following aspects:

(i) *Purity in non-evacuated membrane tank.*
As required by GLACIER to drift the electrons over 20 m, the oxygen equivalent impurity should be kept within 100 ppt level. This is challenging since the membrane tank could not be evacuated since it does not sustain vacuum inside. Alternatively, by means of argon gas "piston-purge" we should be able to remove the air and achieve the required purity following continuously purifying both the gas and liquid. The $3 \times 1 \times 1$ m^3 will be used to test this principle.

(ii) *The large area charge readout system.*
The $3 \times 1 \times 1$ m^3 TPC will be implemented with 12 50×50 cm^2 anode and LEM panels respectively. It will give information on the gain, charge sharing uniformity, signal to noise ratio and so on for such a large area readout system.

(iii) *Accessible cold front-end electronics.*
The $3 \times 1 \times 1$ m^3 TPC will be implemented with 1920 readout channels. These front-end cards will be placed near the argon vapour phase in order to reduce the noise. Furthermore, these cards could be accessed and replaced without polluting the argon in the tank.

Journal of Physics: Conference Series **650** (2015) 012011 doi:10.1088/1742-6596/650/1/012011

(iii) *Cryogenic and purification system.*
For the membrane tank, the rated pressure is very small. Thus the cryogenic system needs to be designed to safely compensate the boiling-off gas without reaching the rating pressure. A scalable boiling-off compensation principle and sub-immersed liquid argon recirculation system will be tested.

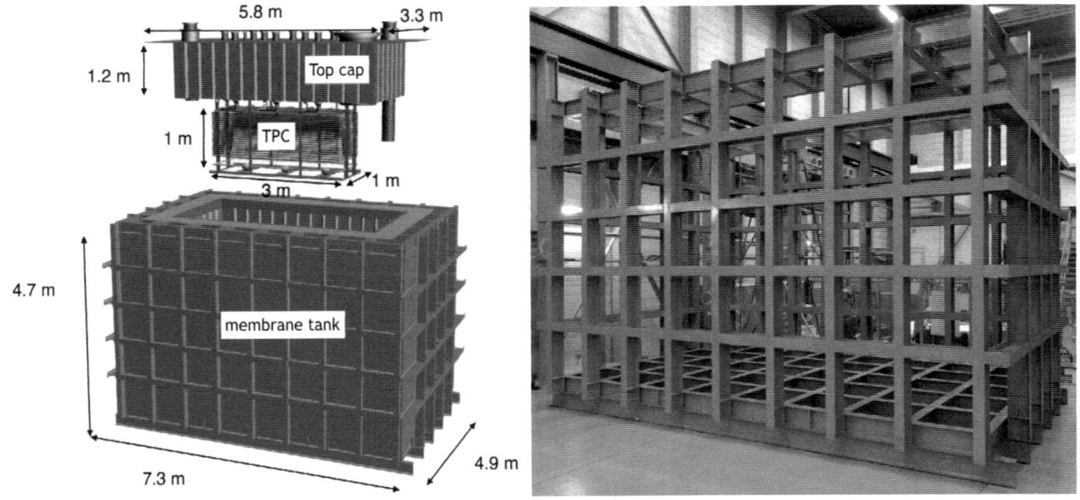

Figure 5. Left: 3D drawings of the $3 \times 1 \times 1$ m^3 TPC, the top cap and the membrane tank; right: the completed outer structure to host the membrane inner tank and the insulation panels.

3.2. The 6×6×6 m^3 demonstrator

In the mean time of building and testing the $3 \times 1 \times 1$ m^3 prototype, the $6 \times 6 \times 6$ m^3 LAr LEM TPC will be built and tested in a beam at the CERN north area. It has a LAr fiducial mass around 300 ton and a total mass around 700 ton. As a technical demonstrator, it will further prove the feasibilities as listed for the $3 \times 1 \times 1$ m^3. Moreover, with a drift length of 6 m, it'll be able to test the image distortion over such a long drift together with the high voltage generation and feedthrough. With a typical drift field of 1 kV/cm, it requires a negative potential of 600 kV at the cathode. How to generate and feed in such a high voltage without discharging against the membrane tank is not clear and thus needs further R&D. This is the way towards GLACIER, since the cathode voltage will be as high as 2 mega-volts for a drift length of 20 m.

Besides the technical goals, the $6 \times 6 \times 6$ m^3 TPC programme also has rich physics topics. It'll be the first double phase LAr TPC in a test beam. Its fiducial mass is big enough to fully contain the hadronic shower from the multi-GeV pion interactions. This allows us to precisely calibrate the energy resolution of liquid argon in an energy range of 0-10 GeV as proposed by LBNO. With a large data set collected, we'll be able to develop and improve the existing particle tracking and identification algorithms. Due to the similarity in the event topology between a neutrino and pion interaction, the data will be used to develop the automatic neutrino event reconstruction algorithm.

The activities of the $6 \times 6 \times 6$ m^3 go in parallel with the $3 \times 1 \times 1$ m^3. A detailed technical design report was drafted [12], and it is approved by CERN with an official experiment name of WA105 [13]. Within an affordable timescale, we are able to build and operate this device before the long shutdown of the LHC in 2018.

Journal of Physics: Conference Series **650** (2015) 012011 doi:10.1088/1742-6596/650/1/012011

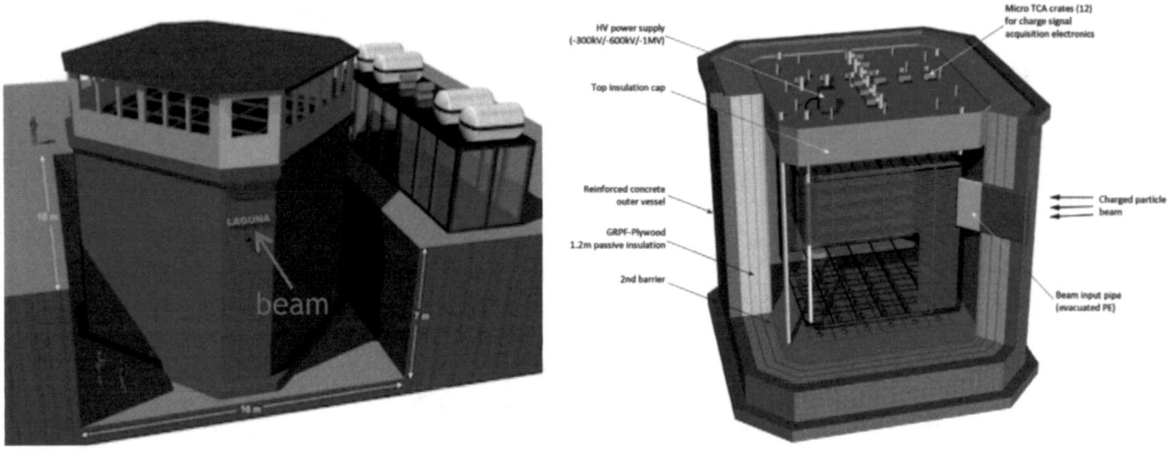

Figure 6. The $6 \times 6 \times 6$ m^3 LAr LEM TPC to be put in a test beam at the CERN north area.

4. Conclusions

The LAr LEM TPC is the state-of-art technology for neutrino detection thanks to its superb tracking and calorimetry performance. The LBNO experiment is competitive in its physics potential to discover the neutrino mass hierarchy and measure the δ_{CP} within an affordable timescale. Good R&D on the charge readout system have been performed in the sense of reducing the anode capacitance, optimising the LEM geometry for a higher gain, simplify the extraction grid and improve the gain uniformity. Having achieved the desired performance, we're moving forward for the $3 \times 1 \times 1$ m^3 and $6 \times 6 \times 6$ m^3 demonstrators to prove the feasibility of the LAr LEM TPC technology to extend to the multi-ton and multi-hundred-ton scales. These demonstrators will bring vital input for the design and operation of the GLACIER 20 and 50 kton for LBNO.

5. Reference

[1] F. P. An *et al.*, "Observation of Electron-Antineutrino Disappearance at Daya Bay" Phys. Rev. Lett. **108**, 171803 (2012).

[2] J. K. Ahn *et al.* "Observation of Reactor Electron Antineutrinos Disappearance in the RENO Experiment" Phys. Rev. Lett. **108**, 191802 (2012).

[3] K. Abe *et al.* "Observation of Electron Neutrino Appearance in a Muon Neutrino Beam" Phys. Rev. Lett. **112**, 061802 (2014).

[4] Stahl. A *et al.* "Expression of interest for a very long baseline neutrino oscillation experiment (LBNO)" SPSC-EOI-007; http://cds.cern.ch/record/1457543?ln=en

[5] S. K. Agarwalla *et al.* "The mass-hierarchy and CP-violation discovery reach of the LBNO long-baseline neutrino experiment" J. High Energy Phys. **05** (2014) 094 [arXiv:1312.6520]

[6] RD51 collaboration, http://rd51-public.web.cern.ch/RD51-Public/

[7] http://laguna.ethz.ch:8080/Plone/deliverables/laguna-lbno-284518-deliverables

[8] A. Badertscher *et al.*, "First operation of a double phase LAr Large Electron Multiplier Time Projection Chamber with a 2D projective readout anode" Nucl. Instrum. Meth. **A641**, 48-57 (2011).

[9] J. Rico, "First study of the stopping Muon sample with the ICARUS T600 detector" Ph.D. Dissertation, ETH 14906 (2002).

[10] C. Cantini *et al.*, "Long-term operation of a double phase LAr LEM Time Projection Chamber with a simplified anode and extraction-grid design" JINST **9**, P03017 (2014) [arXiv:1312.6487 [physics.ins-det]].

[11] C. Cantini *et al.*, "Performance study of the effective gain of Large Electron Multipliers in LAr-LEM TPCs" [arXiv:1412:4402 [physics.ins-det]].

[12] L. Agostino *et al.* "LBNO-DEMO: Large-scale neutrino detector demonstrators for phased performance assessment in view of a long-baseline oscillation experiment" [arXiv:1409.4405]

[13] http://wa105.web.cern.ch/wa105/

Journal of Physics: Conference Series **650** (2015) 012012 doi:10.1088/1742-6596/650/1/012012

Measurement of scintillation and ionization yield with high-pressure gaseous mixtures of Xe and TMA for improved neutrinoless double beta decay and dark matter searches

Y Nakajima, A Goldschmidt, H S Matis, D Nygren, C Oliveira, J Renner

Lawrence Berkeley National Laboratory, Berkeley, CA 94720, USA

E-mail: YNakajima@lbl.gov

Abstract. Liquid Xe TPCs are among the most popular choices for double beta decay and WIMP dark matter searches. Gaseous Xe has intrinsic advantages when compared to Liquid Xe, specifically, tracking capability and better energy resolution for double beta decay searches. The performance of gaseous Xe can be further improved by molecular additives such as trimethylamine(TMA), which are expected to (1) cool down the ionization electrons, (2) convert Xe excitation energy to TMA ionizations through Penning transfer, and (3) produce scintillation and electroluminescence light in a more easily detectable wavelength (300 nm). These features may provide better tracking and energy resolution for double-beta decay searches. They are also expected to enhance columnar recombination for nuclear recoils, which can be used for searches for WIMP dark matter with directional sensitivity. We constructed a test ionization chamber and successfully measured scintillation and ionization yields at high precision with various Xe and TMA mixtures and pressures. We observed the Penning effect and an increase in recombination with the addition of TMA. However, many undesired features for dark matter searches, such as strong suppression of the scintillation light and no sign of recombination light, were also found. This work has been carried out within the context of the NEXT collaboration.

1. Introduction

Xenon TPCs are among the most popular and sensitive detector technologies to search for WIMP dark matter and neutrinoless double beta decay. While liquid Xe is more commonly used for such applications, gaseous Xe has intrinsic advantages over liquid Xe. For neutrinoless double beta decay search, gaseous Xe provides superior energy resolution and tracking capability. Those excellent capabilities are demonstrated in Refs. [1, 2] and being carried out within the NEXT experiment [3].

In addition, there is a potential capability of performing dark matter searches with directional sensitivity utilizing columnar recombination [4, 5]. To put this idea into practice, it is essential to efficiently cool ionized electrons. The directional sensitivity can be further enhanced by the Penning effect, which converts primary excitation into an ionization signal. Hence, doping with molecular additives would be essential to achieve enhancement of columnar recombination.

We are exploring possible performance improvements by adding trimethylamine(TMA) to

Journal of Physics: Conference Series **650** (2015) 012012 doi:10.1088/1742-6596/650/1/012012

gaseous Xe. We report measurements of ionization and scintillation yields with a test ionization chamber that we constructed at Lawrence Berkeley National Laboratory.

2. Performance improvements with TMA

For typical Xe TPCs, observable signals are scintillation light from excited Xe ($Xe \to Xe^*$, $Xe^* + Xe \to Xe_2^* \to 2Xe(g.s.) + h\nu$) and ionized electrons ($Xe \to Xe^+ + e^-$). The relative size of those signals can be altered with the presence of sufficient recombination, $Xe^+ + e^- \to Xe^*$ and $Xe^* + Xe \to Xe_2^* \to 2Xe(g.s.) + h\nu$, which effectively converts ionization signals into the scintillation signals. A sufficient amount of columnar recombination would be required for a dark matter search with directional sensitivity. However, due to large electron diffusion, such recombination in a pure Xe medium is expected to be negligible.

TMA is known to have a large inelastic cross section for electrons due to various vibrational and rotational modes, and provides efficient cooling of ionization electrons, which is essential for the recombination processes. In addition, the ionization potential of TMA (7.85 ± 0.05 eV [6]) is well matched to the energy level of the first excited state of Xe (8.3 eV [7]). This allows a Penning transfer ($Xe^* + TMA \to Xe + TMA^+ + e^-$), which increases the intrinsic amount of ionization that would undergo columnar recombination. We also expect that the charge exchange process, $Xe^+ + TMA \to Xe + TMA^+$, would occur quickly. With efficient Penning and charge-exchange transfers, initial excited and ionized Xe would be largely converted into ionized TMA^+ and electron pairs. Finally, TMA produces fluorescence light at ~ 300 nm, which is much more easily detectable compared to the light from Xe (~ 170 nm) because of typically higher PMT quantum efficiency at 300 nm. If recombination of ionized TMA produces similar fluorescence light, via $TMA^+ + e^- \to TMA^* \to TMA(g.s.) + h\nu(300\ \mathrm{nm})$, then the detection efficiency for columnar recombination would be significantly increased.

In addition, Penning transfer increases the number of ionized electrons, which improves statistical precision of the energy measurement and track reconstruction. Tracking performance would also be improved with TMA by reducing the diffusion of ionization electrons [8]. These features would further benefit neutrinoless double-beta decay searches in gaseous Xe.

3. Experimental setup

In order to test the properties of the gas mixtures of Xe and TMA, especially Penning transfer efficiency and fluorescence light from recombination processes, we built a test ionization chamber as shown in Fig. 1.

A pair of parallel plate electrodes with gap width of 5 mm is placed in the chamber that can hold pressurized gas up to 8 bar. One of the electrodes is separated into a 2.5 cm-diameter inner electrode and an outer electrode. The signal induced in the inner electrode is used for the analysis, in order to avoid the effects from the charge produced outside the gap between the electrodes. Four PMTs are placed at the circumference of the electrodes to detect scintillation light from the gas between the electrodes. A long-pass filter that blocks light below 250 nm is installed in front of one of the PMTs, so that this PMT is only sensitive to the scintillation light from TMA, eliminating the Xe VUV light. We irradiated the gas between the electrodes with $\sim$60 keV gamma rays from a 10 mCi ^{241}Am radioactive source. Currents from electrodes and PMTs were read with pico-ammeters in DC mode. A more detailed description of the setup can be found in Ref. [9].

The gas flow in our system is illustrated in Fig. 2. Once the chamber is filled with the gas mixture, it is sealed and the gas inside is purified by continuous circulation through a SAES MC50-702F room temperature gas filter ("Cold getter" in Fig. 2). While this filter is essential for removing oxygen, which blocks the scintillation light from Xe, we found that the filter also takes a significant amount of TMA from the Xe + TMA mixture. We noticed that the TMA concentration in the gas mixture decreases shortly after introducing the gas into the system, and

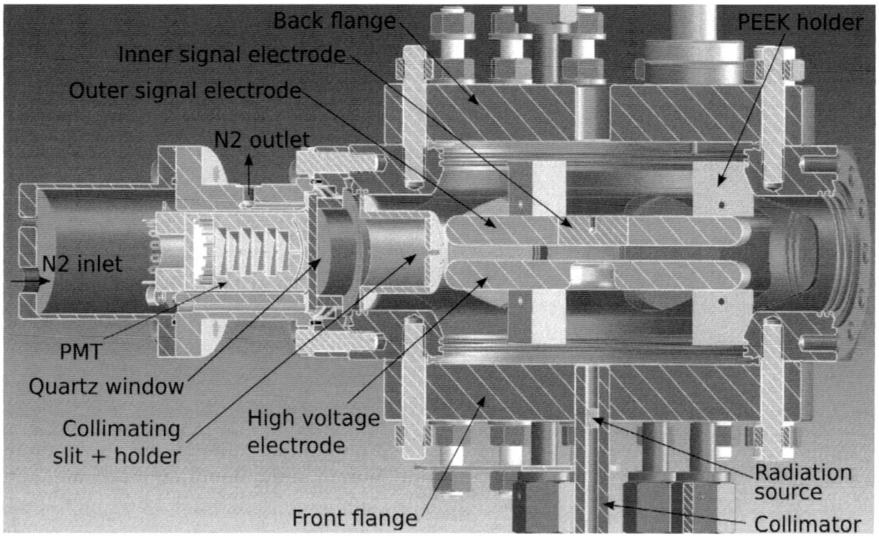

Figure 1. Schematic of the ionization chamber.

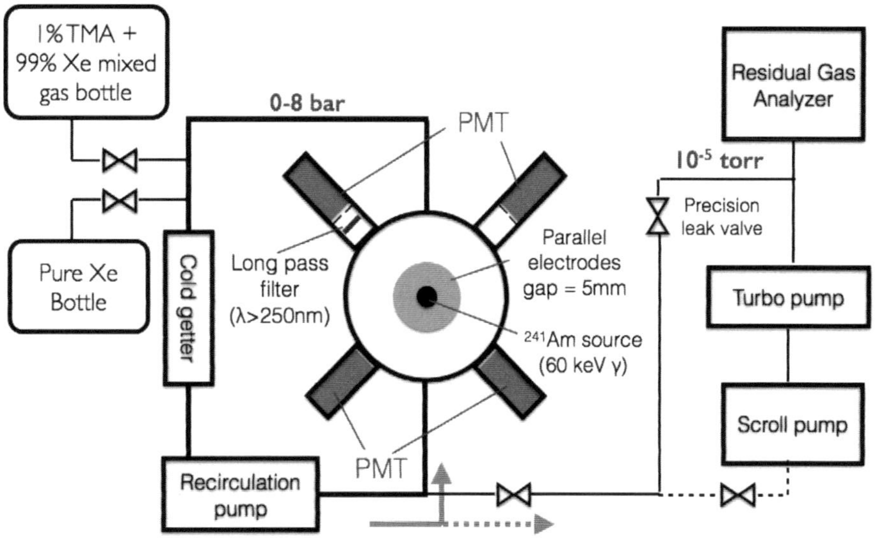

Figure 2. Schematic diagram of the gas flow system.

then stabilizes after about one hour of recirculation. This makes it difficult to precisely control the TMA fraction in the chamber, and hence we continuously monitored the gas composition using a residual gas analyzer (RGA). As shown in Fig. 2, a small fraction of the gas inside is continuously introduced into the RGA through the precision leak valve. The gas pressure around the RGA was adjusted to $\sim 10^{-5}$ torr. Figure 3 shows a typical mass spectrum of the Xe+TMA mixture, obtained with the RGA. Singly and doubly ionized Xe, as well as TMA mass peaks, can be seen. We monitored the relative height of Xe and TMA pressures, and extracted the TMA fraction in the main chamber. The absolute calibration of the TMA fraction was done by introducing 1% TMA and 99% Xe mixture to the RGA directly, without the filter.

Journal of Physics: Conference Series **650** (2015) 012012 doi:10.1088/1742-6596/650/1/012012

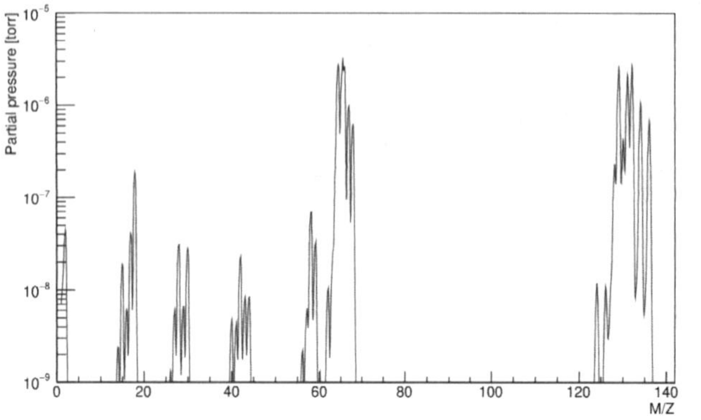

Figure 3. Example output of an RGA analog mass scan for a Xe+TMA gas mixture. The horizontal axis presents the mass divided by the unit charge. Peaks around $M/Z = 130(65)$ correspond to singly(doubly) ionized Xe atoms. Peaks around M/Z of 58, 42 and 30 are mostly from TMA and its fragments. Residual components of H_2(M/Z=2), H_2O(M/Z =18) and N_2(M/Z=28) are also shown.

4. Results with pure Xe

The system was initially tested with pure Xe gas [9]. Figure 4 shows results of light yield measurements at various electric fields and pressures. The amount of energy deposition in the active region, which changes with pressure, is corrected using a GEANT4-based Monte Carlo simulation. Primary scintillation light for $E/\rho < 10^5$ [Vcm^{2}g^{-1}] and proportional amplification at higher electric fields (electroluminescence light) are clearly seen. The results are consistent among the different pressures up to 8 bar. Figure 5 shows the current induced in the electrode due to ionization processes. Again, good agreement between different pressures was seen. The charge yield is relatively flat for $E/\rho > 10^3$ [Vcm^{2}g^{-1}], which indicates an accurate charge correction efficiency at various conditions. In addition, this flatness which extends to higher fields of $E/\rho \approx 10^5$ [Vcm^{2}g^{-1}] suggests that the increase in scintillation yield at the same higher field region is due to proportional amplification without charge amplification (avalanche). The reduction of the charge yield at the low field region of $E/\rho < 10^3$ [Vcm^{2}g^{-1}] is likely due to electrons diffusing out. Those results are in good agreement with the previous measurements [10, 11]. A more detailed description of test results with pure Xe can be found in Ref. [9].

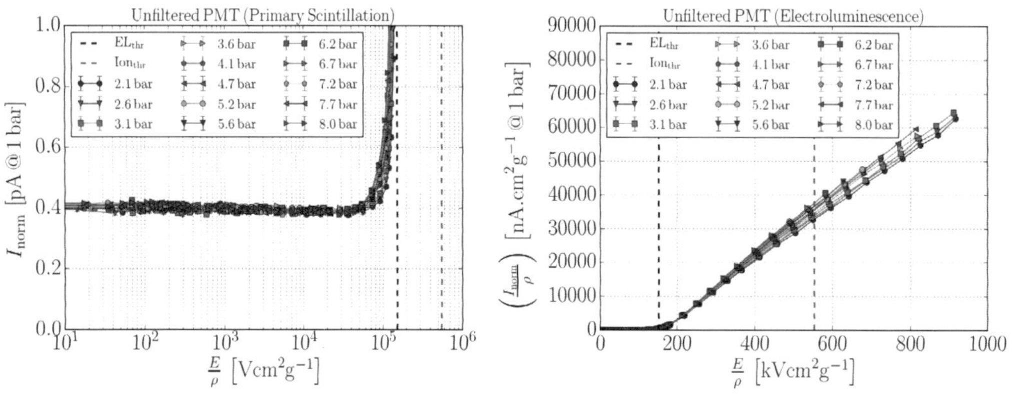

Figure 4. Light yield of pure Xe gas as a function of external electric field for different total pressures. (Left) Results in the primary scintillation region. (Right) Results in the Electroluminescence region.

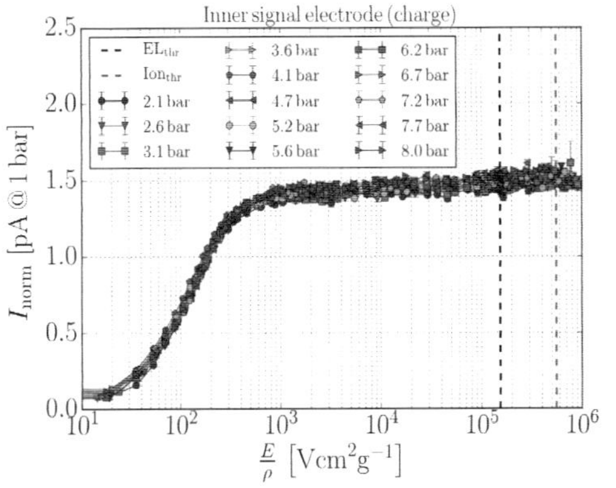

Figure 5. Charge yield of pure Xe gas as a function of external electric field for different total pressures.

5. Results with Xe and TMA mixture

5.1. Charge yield

Figure 6 shows the charge yield observed at the inner electrode at approximately 4 bar total pressure and for various TMA concentrations. We found that the slope of charge yield vs. electric field at $10^3 < E/\rho < 10^5$ (Vcm^2g^{-1}) increases as the TMA fraction increases. That can be interpreted as an effect of increased recombination, since recombination is larger at lower electric field and becomes negligible at high electric field. Therefore, the larger slope indicates larger amount of recombination at a low electric field with more TMA. Furthermore, we observed that the charge yield at a high electric field of $\sim 10^5$ Vcm^2g^{-1} increases as the TMA fraction increases, which we interpreted as the effect of Penning transfer. We also found that charge amplification starts at lower electric field with the addition of TMA. This could be due to lower ionization potential of TMA and also due to Penning transfer.

Figure 7 shows the relative size of charge yield for $10^5 < E/\rho < 2 \times 10^5$ (Vcm^2g^{-1}), as a function of TMA concentration. We found a clear correlation between charge increase and TMA concentration. The charge yield is increased by $\sim 5\%$ with 1% TMA concentration. In gaseous Xe, the relative size of excitation and ionization is measured to be $1/W_{sc} : 1/W_i \sim 1 : 2.5$ [12], where $W_{sc(i)}$ is the amount of energy loss of electrons needed to excite (ionize) one Xe atom. Penning efficiency, $\epsilon(Penning)$, can be calculated using the relative charge yield between Xe+TMA mixture to pure Xe, $I(\text{Xe+TMA})/I(\text{pure Xe})$, as

$$\epsilon(Penning) = \frac{W_{sc}}{W_i}\left(\frac{I(\text{Xe+TMA})}{I(\text{pure Xe})} - 1\right). \tag{1}$$

Hence, a 5% increase in the charge yield translates into 10-15% of Penning transfer efficiency. We observed a clear signature of Penning transfer. However, the amount of Penning transfer is found to be relatively small to make any significant impact on the ionization signal.

5.2. Light yield

Figure 8 shows results of light yield measurements with various TMA concentrations with an unfiltered PMT. We found a large reduction of primary scintillation light once a very small TMA fraction, as low as $\sim 0.01\%$, is introduced. On the other hand, we observed electroluminescence light at higher electric field, as shown in Fig. 9. This EL light is observed for both unfiltered and filtered PMTs, which indicates that it is scintillation light from TMA and the light from Xe is practically lost.

Although the amount of primary scintillation light is largely suppressed with Xe+TMA mixture, we observed non-zero amount of the scintillation light. Figure 10 shows the observed

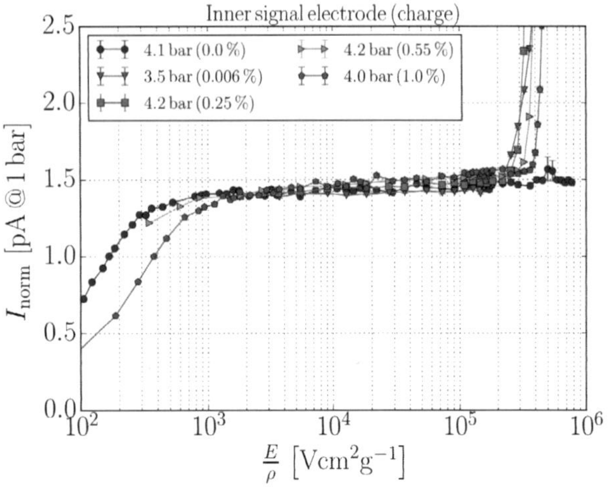

Figure 6. Charge yield of gas mixture of TMA and Xe as a function of external electric field for a total pressure of approximately 4 bar and various TMA concentrations.

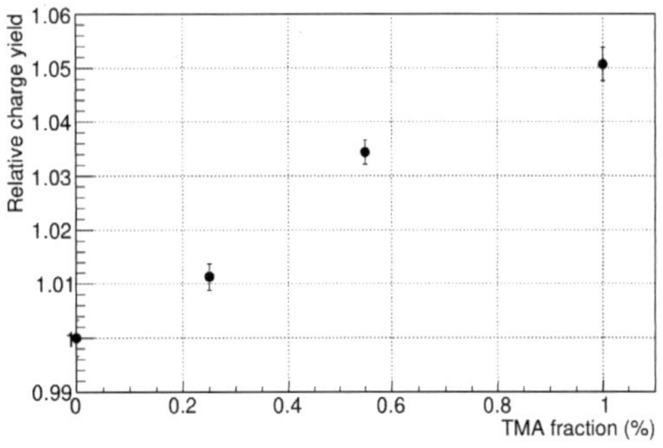

Figure 7. Relative size of averaged charge as a function of TMA concentration for a total pressure of approximately 4 bar. The data for $10^5 < E/\rho < 2 \times 10^5$ ($\mathrm{Vcm^2g^{-1}}$) are used to avoid the effect of the recombination process.

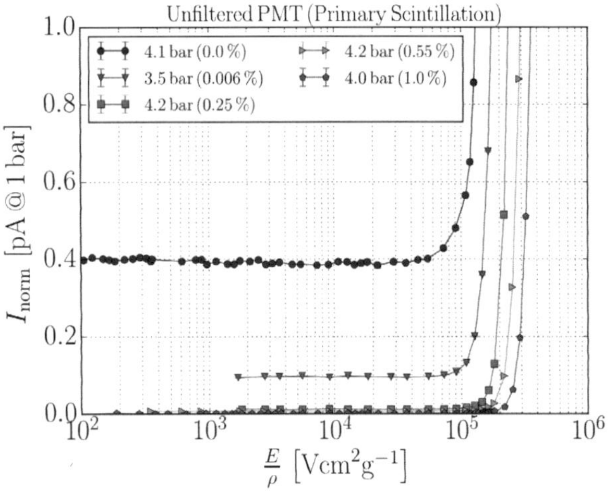

Figure 8. Primary scintillation light yield with Xe+TMA gas mixture, measured at approximately 4 bar total pressure and various TMA concentration.

current from the unfiltered PMT with $\sim 0.5\%$ TMA concentration for different total pressures. The size of the signal is approximately 1-4% of that of pure Xe. This is roughly consistent with the expectation from direct excitation of TMA, after taking into account the difference of PMT quantum efficiency between 170 nm and 300 nm and the difference in excitation cross sections between Xe and TMA. A consistent amount of scintillation light was observed with the filtered

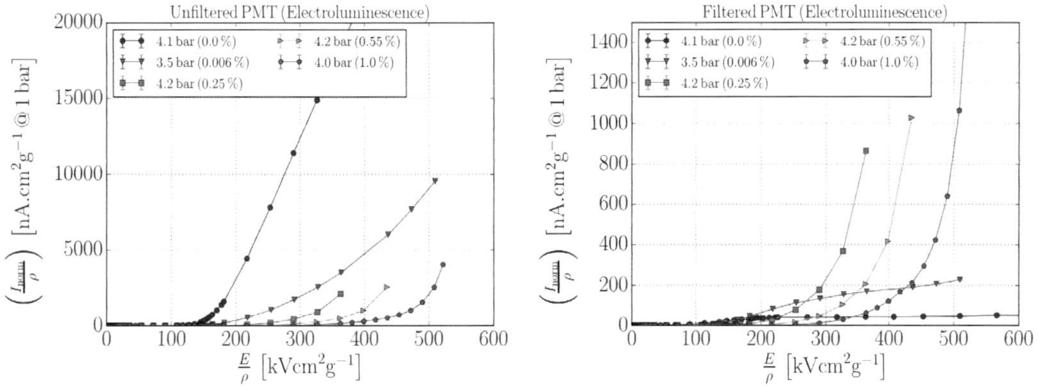

Figure 9. Electroluminescence light yield with Xe+TMA gas mixture, measured at approximately 4 bar total pressure and for various TMA concentrations. The left(right) panel shows the results from unfiltered(filtered) PMT.

PMT as well, which also supports the conclusion that the observed light is only emitted from TMA. The suppression of the light yield at higher pressure is consistent with the self-quenching effect of the TMA [13]. A more careful investigation of this self-quenching effect is underway. If the TMA recombination produces fluorescence light, it would be anti-correlated to the charge yield and would appear as a negative slope of light yield as a function of electric field. However, we observed a constant light yield as a function of electric field for $10^3 < E/\rho < 10^5$ ($\mathrm{Vcm^2g^{-1}}$). Therefore, we conclude that no significant additional scintillation light, such as light from TMA recombination, energy transfer from Xe, is observed.

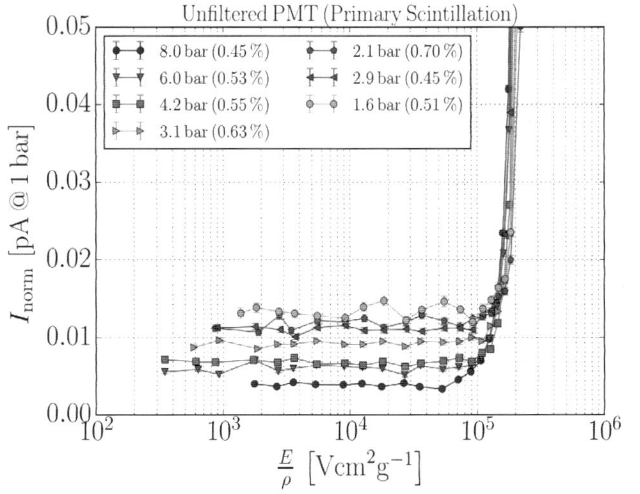

Figure 10. Primary scintillation light yield with Xe+TMA gas mixture. Results of unfiltered PMT at approximately 0.5% TMA concentration and for different total pressures are shown.

6. Discussion

One of the most significant effects that we observed with the Xe and TMA mixture is the high suppression of the primary scintillation light. It is likely that TMA is either absorbing Xe scintillation light, or taking out energy from Xe excited states but not releasing the energy in the form of observable light. We observed enhancement of the recombination process through the reduction of charge yield at low electric field. However, we did not observe a corresponding increase of scintillation light. This can be due to either lack of charge-exchange process ($Xe^+ + TMA \rightarrow Xe + TMA^+$) and also due to lack of fluorescence after TMA recombination. An increase of charge yield due to Penning effect was clearly observed, and Penning efficiency was found to be 10-15% for 1% TMA concentration at 4 bar total pressure. This, however, corresponds to an increase in the number of ionization electrons at the level of only 5% and would not give significant improvement in the detector performance. All of those observations indicate that, unfortunately, TMA is not a suitable additive to enhance performance of gaseous Xe TPC for dark matter searches. While the suppression of scintillation light is also problematic for neutrinoless double beta decay searches, other features of TMA, such as reduction of diffusion and increase of charge yield, can still be beneficial for this application and are actively studied [8, 14].

On the other hand, our system is capable of measuring scintillation and ionization yield for any gas mixture at high precision. Other gas mixtures, such as Xe and Triethylamine(TEA), may be tested to explore ideal gas mixtures.

7. Summary

Using a test parallel-plate ionization chamber, we successfully made high-precision measurement of scintillation and ionization yield for Xe+TMA gas mixture at various electric fields. Although we found the Penning effect and enhancement of the recombination process with the addition of TMA, many undesired features, such as strong suppression of the scintillation light and no sign of recombination light, were also found. Therefore, it is likely difficult to utilize TMA to enhance the performance of gaseous Xe TPCs in dark matter searches. Other features of TMA, such as diffusion reduction and charge gain increase, can still benefit neutrinoless double beta decay searches.

Although TMA is found to be not an ideal additive, our system has proven to be capable of measuring scintillation and ionization yield at high precision. We are exploring the possibility of testing other gas mixtures, such as Xe+TEA.

8. Acknowledgments

We acknowledge the indispensable technical support from Tom Miller (LBNL) during the design, construction and running of the detector. This work was supported by the Director, Office of Science, Office of Basic Energy Sciences, of the US Department of Energy under contract no. DE-AC02-05CH11231. J. Renner (LBNL) acknowledges the support of a US DOE NNSA Stewardship Science Graduate Fellowship under contract no. DE-FC52-08NA28752.

References

[1] Alvarez V *et al.* (NEXT Collaboratoin) 2013 *Nucl.Instrum.Meth.* **A708** 101–114
[2] Alvarez V *et al.* (NEXT Collaboratoin) 2013 *JINST* **8** P09011
[3] Granena F *et al.* (NEXT Collaboratoin) 2009 *Preprint* arXiv:0907.4054
[4] Nygren D 2013 *J.Phys.Conf.Ser.* **460** 012006
[5] Nakajima Y *et al. In these proceedings*
[6] NIST Chemistry WebBook URL `http://webbook.nist.gov/cgi/inchi/InChI%3D1S/C3H9N/c1-4(2)3/h1-3H3`
[7] NIST Atomic Spectra Database URL `http://www.nist.gov/pml/data/asd.cfm`
[8] Gonzalez-Diaz D, Alvarez V, Borges F, Camargo M, Carcel S *et al.* 2015 *Preprint* arXiv:1504.03678
[9] Oliveira C, Gehman V, Goldschmidt A, Nygren D and Renner J 2015 *Physics Procedia* **61** 742 – 749
[10] Monteiro C, Fernandes L, Lopes J, Coelho L, Veloso J *et al.* 2007 *JINST* **2** P05001
[11] Freitas E, Monteiro C, Ball M, Gomez-Cadenas J, Lopes J *et al.* 2010 *Phys.Lett.* **B684** 205–210
[12] Renner J *et al.* (NEXT Collaboration) 2014 *Preprint* arXiv:1409.2853
[13] Cureton C G, Hara K, O'Connor D V and Phillips D 1981 *Chemical Physics* **63** 31–49
[14] Alvarez V *et al.* (NEXT Collaboration) 2014 *JINST* **9** C04015

Lessons from the operation of the 'Penning-Fluorescent' TPC and prospects

Diego Gonzalez-Diaz, F. Aznar, J. Castel, S. Cebrián, T. Dafni, J. A. García, J. G. Garza, H. Gómez, D. C. Herrera, F. J. Iguaz, I. G. Irastorza, A. Lagraba, G. Luzón, A. Rodríguez, E. Ruiz-Choliz, L. Segui, A. Tomás

Laboratorio de Física Nuclear y Astropartículas, Universidad de Zaragoza, Zaragoza, Spain

E-mail: `diegogon@cern.ch`

E. Ferrer-Ribas, I. Giomataris

IRFU, Centre d'Études Nucléaires (CEA), Saclay, France

Abstract. We have recently reported the development of a new type of high-pressure Xenon time projection chamber operated with an ultra-low diffusion mixture and that simultaneously displays Penning effect and fluorescence in the near-visible region (300 nm). The concept, dubbed 'Penning-Fluorescent' TPC, allows the simultaneous reconstruction of primary charge and scintillation with high topological and calorimetric fidelity.

1. Introduction and main results in charge mode

The possibility of using 'Penning-Fluorescent' gas mixtures was addressed as early as at least 1981, with the seminal work of A. Policarpo on light multiplication [1], and coetaneous to the invention of the TPC by D. Nygren and J. Marx [2]. The possibility of the confluence of the two techniques was brought forward recently, by Nygren himself [3], in the context of novel gas mixtures for $\beta\beta0\nu$-searches, for which the TPC technology must be desirably pushed beyond the current technological limits both in the calorimetric and topological front. We have extensively studied the possibility of using Xenon/trimethylamine Penning-Fluorescent mixtures in a series of papers [4, 5, 6, 7, 8] culminating in [9], and the main lessons there extracted are elaborated further in this proceeding. A summary of the impact of the TMA addition on the primary charge and scintillation of gaseous Xenon is summarized in Fig.1, while the topological and calorimetric properties obtained inside a 1.1 kg fiducial mass (as recorded with microbulk Micromegas) are given in Fig.2. An extensive description of the chamber performance for the reconstruction of γ-tracks in charge mode can be found in [9], and for the light properties the reader is referred to the work of Y. Nakajima also presented at this conference [10]. A microscopic modelling of the Micromegas response is in preparation [11].

Importantly, it must be noted at the outset that the term 'Penning-Fluorescent' refers here to wavelength-shifted fluorescence, i.e., to scintillation with a characteristic spectrum centered around the visible range (hence, noble gas VUV-fluorescence must be necessarily quenched by the mixture too). In this shifted band, the parasitic photo-effect at metallic surfaces as well as photon transmission become extremely benign, facilitating both the detection of the primary scintillation with conventional PMTs and charge detection with micro-pattern gaseous detectors.

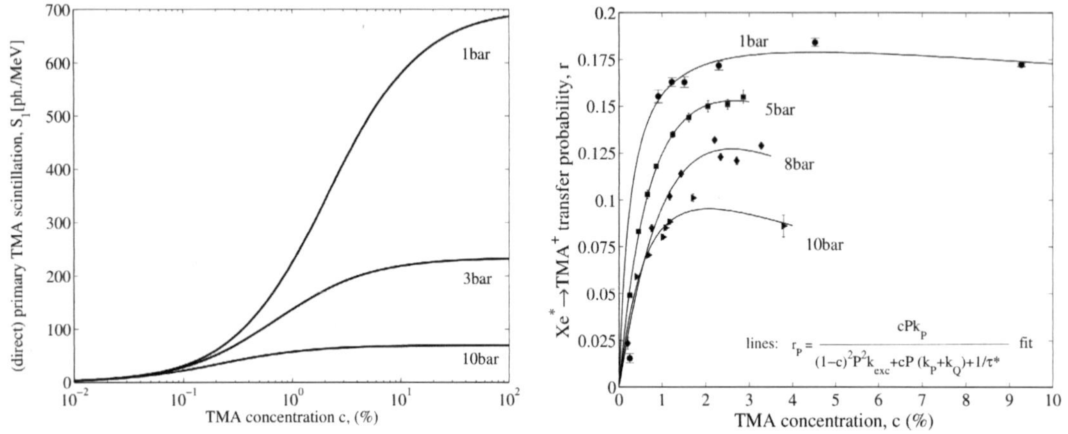

Figure 1. Left: estimated primary scintillation yields (S_1) in the TMA fluorescence band (300 nm) for Xe-TMA mixtures under the conservative assumption that only direct TMA excitation produces observable photons. Right: Penning transfer rates obtained from a first-order modelling of the gain observed in a Micromegas charge amplification device, assimilated to a parallel plate chamber. A detailed 3D field calculation yields maximum transfer rates that are higher by up to 50%, and will be published elsewhere [11].

1.1. Fluorescence

The TMA fluorescence spectrum is centered at around 300 nm and ends at about the work function of copper, while TMA itself is strongly opaque to Xenon light. A lower bound to the primary scintillation expected from a Xenon-TMA mixture can be derived from studies at low pressure, and clearly requires confirmation at the pressures of interest for $\beta\beta0\nu$: omitting conservatively the presence of Xe*, Xe_2^* →TMA*(3s) transfers and hence assuming only the processes of direct TMA excitation/scintillation and self-quenching [12] and neglecting thus the quenching of TMA excited states by Xenon species (following [13]) such a bound results in:

$$S_1(TMA) = \frac{1}{W_{sc}} \frac{c}{\tau_{3s}(P \cdot c \cdot k_{SQ} + 1/\tau_{3s})} \tag{1}$$

as a function of the concentration (c) and pressure (P). Predicted S_1 yields from direct TMA excitation are shown for illustrative purposes in Fig. 1, for 1, 3 and 10 bar. It has been assumed for TMA a typical value $W_{sc} = 30 \pm 10$ eV ($\simeq 5.5 \times E_{3s}$), the measured lifetime of the 3s state ($\tau_{3s} = 44$ ns) and the TMA self-quenching rate ($k_{SQ} = 1.08$ ns^{-1} bar^{-1}). Additional scintillation hangs on the ability of TMA to quickly dissipate, collisionally, a fairly large amount of internal excess energy (~ 2 eV) following the Xe*, Xe_2^* →TMA* transfers; relaxation to the 3s state may in this way happen while avoiding the dissociative pathways, [13]. S_1 yields as those shown in Fig. 1-left are not hopeless for $\beta\beta0\nu$-experiments or γ-detection in general but certainly require of dedicated detection schemes. A comprehensive experimental survey of the primary and secondary scintillation processes is currently under preparation and preliminary results can be found in [10]. Although that work points in fact to the smallness of the fluorescent transfer rate in Xe-TMA mixtures, the observed scintillation in the TMA band is compatible with the lower bounds given in Fig. 1.

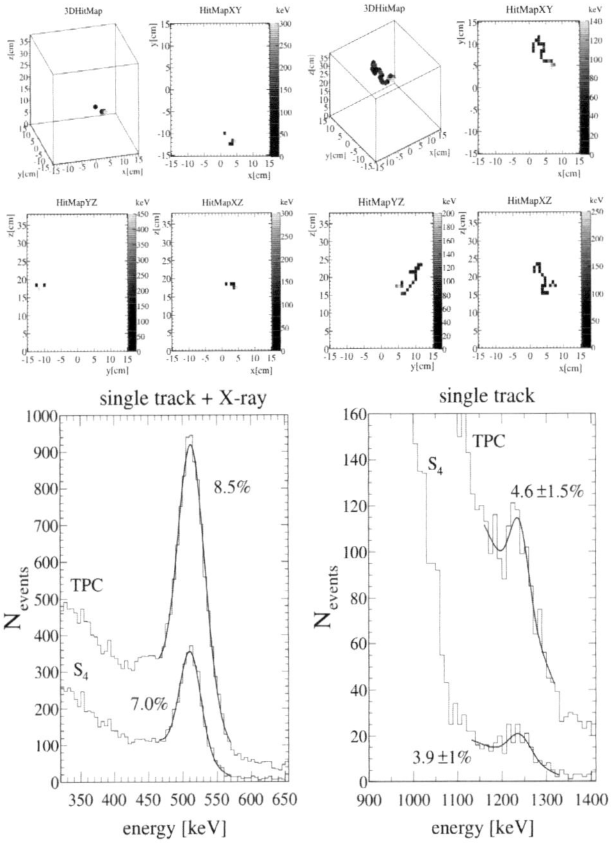

Figure 2. Up: 3D reconstruction and 2D projections of typical 0.511 MeV (left) and 1.275 MeV $\simeq Q_{\beta\beta}/2$ (right) tracks, obtained in the fiducial volume (1.1kg) of the large technological demonstrator NEXT-MM, shown with a voxelization of 8 mm × 8 mm × 10 mm (x, y, z). Down: calorimetric response for each case, indicating FWHM values. The energy resolution for 30 keV X-rays is 14.6%(FWHM).

1.2. Penning transfer rate

To the extent that energy transfers occur with the Xe* states, Penning reactions are energetically viable:

$$\text{Xe}^* + \text{TMA} \rightarrow \text{Xe} + \text{TMA}^+ + \text{e}^- \tag{2}$$

The presence of reactions of the type (2) in Xe-TMA mixtures has been indirectly inferred from the strong gain increase at constant field observed in charge amplification structures ([14], [4]), with an optimum admixture of about 1% TMA for the case of Micromegas. In order to quantify the effect, we present here a simplified analysis performed by assimilating the Micromegas to a parallel plate configuration and using the microscopic code Magboltz for modelling the charge amplification process [15]. An analogous procedure has been used in an earlier work for describing Argon Penning-mixtures in proportional counters [16].

Under the common simplifying assumption that the Penning transfer probability (hereafter r_P) is largely field-independent, a universal value can be extracted for it, as shown in Fig.1-right. A phenomenological description of the trends can be achieved in this case by considering excimer formation (k_{exc}), Penning (k_P) and quenching (k_Q) rates, and an effective life-time of the state from which Penning takes place (τ^*), resulting in the relation:

$$r_P = \frac{cPk_P}{(1-c)^2P^2k_{exc} + cP(k_P + k_Q) + 1/\tau^*(P)} \tag{3}$$

The observed reduction of transfer probability with pressure would seemingly indicate that Penning-transfer from the excimer states is disfavoured and so excimer formation effectively reduces the chances of the transfer to take place.

A more realistic analysis done by including a 3D field calculation will be presented elsewhere [11], although neither it substantially alters the trends nor the conclusions. The maximum transfer rate observed in such an analysis exceeds by 30-50% (relative) the values given here, since the parallel plate approximation overestimates the amplification field.

2. Discussion

2.1. T_o-less Micromegas-only TPC

The energy resolution obtained for 0.511 MeV tracks with Micromegas is at least ×2 better than the one obtained earlier with wire chambers in [17], with Xe-TMA reducing electron diffusion in factors of $\simeq 2$ in all 3 space dimensions relative to, e.g., Xe-CH_4 (Fig. 3) and operation taking place at twice the pressure, thus increasing event containment. Since as of today microbulk Micromegas show good prospects for high radiopurity levels, it may seem that re-sizing the present concept 'as is' could already improve on earlier limits obtained for $\beta\beta0\nu$-decay in high pressure Xenon in [17]. In this scenario, the optimum pressure is undoubtedly a crucial variable to be considered: the maximum gain reported is dramatically dependent on it, e.g. gains of 2000 were shown for small wafers at 5 bar in [4] as compared to 400 at 10 bar, with 200 finally achieved in the system. Any future development would certainly have to address the interplay between signal to noise ratio (S/N) and operating pressure. These type of TPC configurations would constitute excellent X-ray or γ-ray polarimeters, provided the T_o information is not essential [22].[1]

2.2. Micromegas TPC with T_o

In $\beta\beta0\nu$ experiments the determination of the initial scintillation provides the time of the interaction (T_o) allowing event fiducialization and background suppression, as well as enhancing topological reconstruction due to the dependence of diffusion with z. Barium-tagging schemes usually take its knowledge for granted [18], [19], [20].

The level of direct TMA scintillation at 10 bar amounts to a modest yield of $S_1(\sim300\text{nm}) = 58$ ph./MeV, increasing in approximate linear fashion with decreasing pressure. As said, although additional energy transfer from excited Xenon species to the TMA fluorescent 3s-state may happen to some extent, it certainly does not represent a strong effect[10]. Taking such a conservative scintillating figure and assuming a 25% quantum efficiency, nearly a 0.3 solid angle coverage would be needed in order to collect 10 photons for an event at the $Q_{\beta\beta}$ energy, that is certainly unrealistic to be done by sheer increase of the number of PMTs. One may recycle for this purpose a recent idea brought forward in [21], benefiting from the fact that the TMA

[1] The possibility of obtaining the T_o information from the physics of the electron diffusion (pulse width-z correlation) should not be discarded and has been shown earlier at 1bar for localized X-ray irradiation [8]. To the extent that enough S/N and sampling frequency are provided, and the information from displaced X-rays (when present) and/or the combined information from the whole track can be used, this method will find application. Since the low diffusion presumably required for the topological identification and suppression of γ backgrounds is contrary to the high diffusion required for a achieving a strong width-z correlation, the necessity of a dedicated optimization of the sampling frequency, signal shaping and buffer size is likely to ensue upon close examination of this possibility.

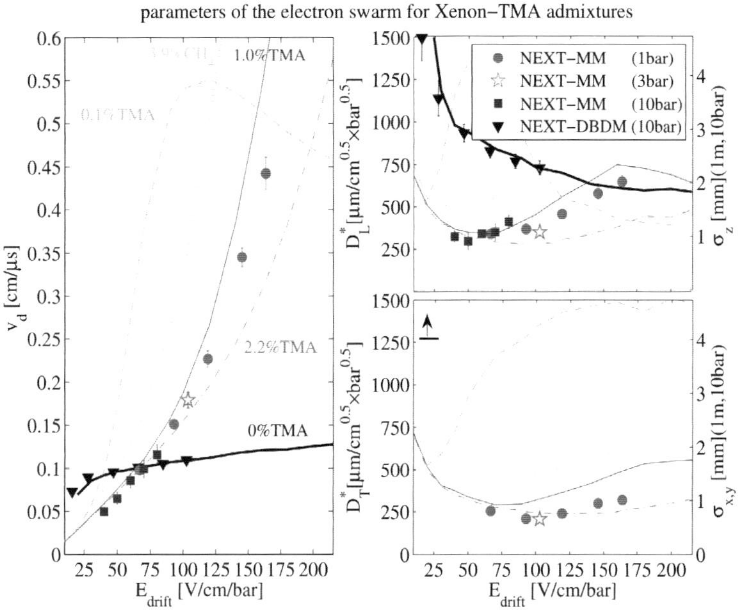

Figure 3. Compilation of the main properties of high pressure Xenon TPCs for 1(1.0), 3(2.7) and 10(10.1) bar and different mixtures, together with simulation (Magboltz 10.0.1). Left: drift velocities. Right: longitudinal and transverse diffusion coefficients. Simulation results for Xe-CH_4 (Gotthard) and Xe-TMA in a 0.1% admixture are also shown for illustration.

emission band matches the peak of the excitation spectrum for common commercial wavelength-shifting (wls) plastics. Although challenging at present, this possibility would allow a drastic reduction in the number of PMTs with additional benefits such as reduced cost and simplicity.

On the other hand such a TPC, with the interior of the field cage and the cathode back covered with wls-plastics would generically constitute an excellent X- and γ- ray detector with good prospects both as Compton camera and polarimeter. With optimized thickness, the surrounding plastics would simultaneously allow vetoing ionizing particles.

2.3. Micromegas+EL TPC with T_o

Despite the large improvement as compared to earlier approaches based on wires, the present work shows that calorimetry with charge readouts still is, as of today, more prone to instrumental limitations than an electroluminescent (EL) approach, particularly at high pressure. This represents an important practical consideration for high precision discovery-type experiments. On the other hand, strategies concerning noise and cross-talk mitigation have been discussed in [9], and a factor $\times 2$ improvement down to 1.5% energy resolution (FWHM) at the energy scale of the ^{136}Xe $Q_{\beta\beta}$ seems realistic without altering the Micromegas readout characteristics. The already modest rate of pixel damage (1%/year) can be brought further down to the 1%-level during realistic experiment live-times through the increase of the pixel granularity and the introduction of tighter quality control (QC) procedures.

Importantly, as compared to light-based readouts, charge amplification and readout with micro-pattern gaseous detectors can more easily approach a 'true-3D' reconstruction (i.e., with the voxelization stemming from pixelization and signal sampling being at the scale at which the relevant track structures emerge: $\sim$ 1mm$\times$1mm$\times$1mm at 10 bar [9]). So, based on the above paragraph it seems worth considering the possibility of a separate topological/calorimetric function, where both tasks can be optimized independently to a large degree. Such a speculative idea could be implemented through a Micromegas readout placed immediately behind the EL region with its cathode serving as the anode of the electroluminescence gap. Calorimetry can be performed with a PMT plane. This brings a number of considerations:

(i) The large feedback and low gains reported in earlier measurements for Micromegas operated in pure Xenon [23] suggest that the use of a VUV-quencher is at least convenient (if not mandatory). At the m-scale Xe/TMA dumps virtually all VUV light, and yet it shows fluorescence in the TMA scintillating band, although special detection schemes are needed (subsection 2.2).

(ii) Quantitatively, it is unclear at the moment what the global effect of the addition of TMA in the calorimetric response is. In some admixtures it can conceivably reduce the electroluminescence threshold and Fano factor in virtue of its scintillating and Penning properties, however it may also increase the intrinsic fluctuations during the production of secondary light, specially if Penning from Xe^* states would be unavoidable during the excitation of the TMA 3s (scintillating) states, hence yielding the characteristic non-linear light-multiplication trends reported in [10]. As a desirable advantage, the mixture will be transparent to the TMA scintillation at the scales of interest, and the nearly 4π coverage enforced by the necessary S_1 sensitivity will guarantee that only the intrinsic light fluctuations will be present, without additional fluctuations stemming from finite photon statistics.

(iii) Due to the relatively high EL-fields, electron transmission from the EL region to the Micromegas amplification holes will approach the optical transparency. This is as low as 12.5% in the present case, although it easily doubles in a triangular pattern [4]. Since, in this approach, Micromegas do not perform the calorimetric function this is not very critical, however the value should be increased to approach at least 50% levels. Standard Micromegas manufactured in the bulk, provide easily higher levels of optical transparency.

(iv) Photo-cathode efficiencies of bare metals in the Xenon band are of the order of 10^{-4} [24]. Further to that, extraction efficiencies are extremely poor in Xenon ($\lesssim 10\%$), [25]. For a typical optical gain of 200 ph./e/cm/bar this will reduce the contribution from parasitic photo-electrons to 1%, and possibly much less since TMA fluorescence sits below the work function of copper and Xenon light will be strongly suppressed. So, stability of Micromegas on such a 'photon-shower' seems to be reasonably possible.

(v) At the Micromegas, light production during the multiplication process will add fluctuations to the reconstructed light. This needs to be assessed experimentally for Xe/TMA admixtures and cannot be discarded out of hand. There is however an important solid angle reduction in the light emerging from microbulk structures, that implies for pure Xenon a modest 20% contribution to electroluminescence light, as reported in [23]. A compromise between high optical transmission and low scintillation from the Micromegas structure will be needed.

(vi) The Micromegas plane will be exposed to sparks from the EL-region, that will likely result in permanent damage, and potentially imply a catastrophic event. It requires special attention and its stability should be demonstrated.

(vii) From the above points it is clear that the incorporation of auxiliary micro-patterned structures for extraction from the EL-region and for pre-amplification (e.g., GEMs or grids) can easily alleviate the most stringent requirements, implying at the same time little impact on the tracking capabilities at the mm-scale.

3. Conclusions

After a series of works performed over the last 3 years, TMA becomes established as an adequate additive for the construction of large Time Projection Chambers, with a number of remarkable properties that make it unique. It is a good Xenon VUV-quencher, yet demonstrating sufficient primary scintillation for MeV-scale track reconstruction and thus simultaneously allowing stable operation of charge amplification micro-pattern devices, by virtue of its wavelength-shifted emission at about 300 nm. It shows Penning effect, therefore enhancing the ionization response, reducing the field needed for gas amplification and potentially improving the intrinsic energy resolution both for primary and secondary ionization, if all spurious instrumental limitations are conveniently mitigated. TMA can be counted amongst the strongest coolants employed in TPCs, comparable to CO_2, CF_4 or DME, and indeed the TPC here introduced shows the smallest electron diffusion of any TPC that is known to authors, at the scale of 1 mm per 1 m drift. TMA has a convenient vapor pressure, exceeding 1bar at ambient temperature, and it can be incorporated to re-circulating systems even in the presence of purifiers. Toxicity and flammability hazards benefit from its strong odour (easily detectable in ppb concentrations), hence posing yet stronger practical limits to the overall gas tightness, limits that were nevertheless comfortably achieved during operation of the TPC whose performance has been summarized here.

Although presently limited in the scintillation and Penning yields, the introduced system instantiates the concept of 'Penning-Fluorescent' TPC, enabling i) $\times 10$ increased track sharpness as compared to pure Xenon, ii) wavelength-shifted primary scintillation and iii) stability of charge amplification based on micro-pattern readouts. The anticipated reduction on the Fano factor could not be demonstrated due to the presence of systematic instrumental limitations in the energy resolution determination. Further research along this promising line will continue.

4. Acknowledgments

The author deeply acknowledges countless discussions with colleagues of the RD51 and NEXT collaborations. Calculations of Penning transfer rates were performed by Ozkan Sahin. The excellent technical support and service from the CERN detector workshop and Servicio General de Apoyo a la Investigacion-SAI of the University of Zaragoza was essential to our investigation. This work was funded by Spanish Ministry of Economy and Competitiveness (MINECO) under grants Consolider-Ingenio 2010 CSD2008-0037 (CUP) and CSD2007-00042 (CPAN), FPA2011-24058, FPA2013-41085, FPA2008-03456, FPA2009-13697, and by the T-REX Starting Grant ERC-2009-StG-240054 of the IDEAS program of the 7th EU Framework Program. Part of these grants are funded by the European Regional Development Plan (ERDF/FEDER).

References

[1] A. J. P. L. Policarpo, Physica Scripta, 23(1981)539.
[2] J. Marx, D. Nygren, Phys. Today 31(10) (1978) 46.
[3] Nygren D., J. Phys. Conf. Ser., 309(2011)012006.
[4] S. Cebrian et al., JINST 8 (2013) P01012.
[5] D. C. Herrera et al., J. Phys. Conf. Ser., **460**(2013)012012.
[6] D. C. Herrera et al., PoS(TIPP2014)057.
[7] V. Alvarez et al. (NEXT Collaboration), JINST 9(2014)P01301.
[8] V. Alvarez et al. (NEXT Collaboration), JINST 9(2014)C04015.
[9] D. Gonzalez-Diaz et al, (NEXT Collaboration), *Accurate γ and MeV-electron track reconstruction with an ultra-low diffusion Xenon/TMA TPC at 10 atmospheres*, arXiv:1504.03678[physics.ins-det], submitted to NIM A.
[10] Y. Nakajima, *Measurement of scintillation and ionization yield with high-pressure gaseous mixtures of Xe and TMA for improved neutrinoless double beta decay and dark matter searches*, arXiv:1505.03585[physics.ins-det].
[11] E. Ruiz-Choliz et al., *Modelling the behaviour of microbulk Micromegas in Xenon/trimethylamine gas*, in preparation.

[12] Clifford G. Cureton et al., Chem. Phys. 63(1981)31.
[13] Kinichi Obi and Yutaka Matsumi, Chem. Phys. 49(1980)95.
[14] B. D. Ramsey, P. C. Agrawal, Nucl. Instr. Meth., A 278(1989)576.
[15] S. F. Biagi., http://magboltz.web.cern.ch/magboltz/.
[16] O. Sahin et al., JINST 5(2010)P05002.
[17] R. Luescher et al., Phys. Lett. B, 434(1998)407.
[18] J. J. Gomez-Cadenas, *The NEXT experiment* arXiv:1411.2433[physics.ins-det], submitted to Nucl. Phys. B(Proc. Suppl.).
[19] D. Sinclair et al., J. Phys. Conf. Ser. 309(2011)012005.
[20] D. R. Nygren *The Biochemistry of Xenon*, presented at the 7th symposium for low-energy rare event detection.
[21] D. R. Nygren, J. Phys. Conf. Ser. 460(2013)012006.
[22] D. Bernard, Nucl. Instr. Meth. A, 729(2013)765.
[23] C. Balan et al. JINST 6(2011)P02006.
[24] J. Derre, Y. Giomataris et al., Nucl. Instr. Meth. A, 449(2000)314.
[25] D. S. Covita et al., Phys. Lett. B, 701(2011)151.

Journal of Physics: Conference Series **650** (2015) 012014 doi:10.1088/1742-6596/650/1/012014

Study of neutrino interactions at the T2K near detector

A. Hillairet

Dept. of Physics and Astronomy, University of Victoria, 3800 Finnerty Rd, Victoria, BC, V8P 5C2 CANADA

Abstract. The Tokai to Kamioka experiment (T2K) is a long baseline neutrino accelerator operating in Japan and designed to measure electron neutrino appearance and muon neutrino disappearance. Three gaseous TPCs compose its near detector complex. The latest T2K analysis combines the results from the appearance and disappearance channels to extract simultaneously four neutrino oscillation parameters. Improvements to the TPC field distortion correction are under development to reduce systematic uncertainties for future oscillation analyses. In parallel a cross section measurement of neutrino-gas interactions happening in the T2K TPCs is in preparation in particular to extract unique proton multiplicity and kinematic information.

1. Introduction

Neutrino physics quite an exciting field since the discovery of neutrino flavor oscillations 20 years ago. Measuring and understanding these oscillation is already an interesting topic by itself but the flavor oscillations imply that the mass eigenstates and flavor eigenstates are different. This is in contradiction with the Standard Model (SM) which was built on the assumption that neutrinos are massless therefore we have first hints of physics beyond the SM. Furthermore, CP symmetry could be violated by neutrino oscillation. A confirmed CP violation in the neutrino sector would open the possibility of using leptogenesis to explain the matter-antimatter in the Universe.

This has motivated the measurement of the mixing angles θ_{12}, θ_{13}, and θ_{23} defining the mixing of the three mass states as well as the difference Δm^2_{21} and Δm^2_{32} between the mass states. The discovery of a relatively large value of θ_{13} has started the era of precision measurements of these parameters. Precision measurements will require a reduction of the neutrino cross section systematic uncertainties and therefore a better understanding of the neutrino-nucleus interactions.

The Tokai to Kamioka (T2K) experiment started operating when θ_{13} had only an upper limit set from previous experiments. T2K was designed to measure θ_{23} and Δm^2_{32} from muon neutrino disappearance and θ_{13} from electron neutrino appearance. The collaboration made the unique choice in accelerator neutrino experiments to use three gaseous TPC in its near detector to reconstruct the momentum and the charge of the particles produced by neutrino interactions. This paper will present the latest T2K results and some overview of work done to understand electric field distortions in the T2K TPCs. The prospects of the first ever analysis of interactions between neutrinos and a gas inside the T2K TPCs will be also presented.

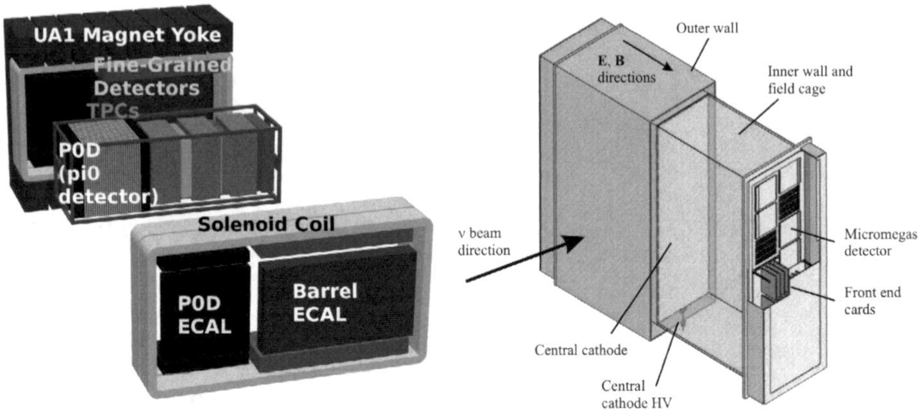

(a) Schematic of ND280 in the opened magnet position.

(b) T2K TPC

Figure 1. . The ND280 schematic shows the tracker with its 3 TPCs and 2 FGDs represented in fake color for display purposes. The walls of the inner box are visible in this TPC cutaway.

2. The apparatus

In T2K, a beam of muon neutrinos is produced at the J-PARC facility in Tokai, Ibaraki, Japan, and it is measured after oscillation 295 km away from the production target by the 50 kT water Cherenkov detector Super-Kamiokande (SK) [1]. T2K was the first accelerator neutrino experiment to adopt an "off-axis" configuration by positioning SK 2.5° away from the beam center. This configuration enhances the neutrino oscillation probability by narrowing the neutrino energy distribution around 600 MeV at the oscillation maximum. The near detector ND280 located 280 m from the production target is also installed "off-axis" and measures the neutrino interaction rate before the oscillation to constrain beam and cross section systematic uncertainties.

ND280 is a complex of multiple subdetectors installed inside the refurbished magnet from the UA1 experiment, which provides a 0.2 T magnetic field (Fig. 1(a)). The central part of ND280 is the tracker which is composed of two fine-grained detectors (FGDs) [3] and three time projection chambers (TPCs) [2]. The FGDs are composed of alternating vertical and horizontal layers of 1 cm^2 square extruded polystyrene scintillator bars read out by wavelength-shifting fibers. The purpose of the FGDs is to act as active targets and provide detailed vertex information of the neutrino interactions. FGD1 is composed entirely of scintillator layers while FGD2 contains scintillator and water layers in order to compare the neutrino interaction rate on carbon and on oxygen. FGD1 and FGD2 are located respectively between TPC1 and TPC2, and TPC2 and TPC3.

The T2K TPCs are composed of an inner box used for the tracking and an outer box fill with $C0_2$ for electric insulation of the inner box (Fig. 2(b)). Each inner box is filled with an argon, CF_4, and isobutane gas mixture at respectively 95, 3, and 2% and uses MicroMegas detectors (MMs) for gas amplification with pad readout. A central cathode splits each TPC into two drift volumes, each equipped with 12 MMs composed of 36×48 pads with a size of 9.8mm $\times$ 7.0mm and an amplification gap of 128 μm. The readout electronic is based on ASIC chips with 72 channels and operating at 25 MHz. The TPCs are used to reconstruct the charged particle's momentum with an inverse momentum resolution of 0.1 $(\mathrm{GeV}/c)^{-1}$. The TPCs are also capable of particle identification using the energy loss to distinguish in particular muons from electrons with a misidentification probability of $< 0.9\%$ from 200 MeV/c to 1.8 GeV/c.

ND280 also contains, upstream of the tracker, the P0D which is dedicated to π^0 reconstruction and made of scintillator bars interleaved with lead and brass sheets [4]. The tracker and the P0D are surrounded by electromagnetic calorimeters (ECals) composed of layers of plastic scintillator bars with lead sheets in between [5].

3. T2K recent results

Before even collecting 10% of its targeted total amount of data, T2K has achieved significant milestones in its scientific program. T2K has performed the first conclusive observation of a neutrino flavour appearance by measuring the appearance of electron neutrinos in SK in the J-Parc muon neutrino beam with a significance of 7.3σ [6]. T2K also produced some of the most precise measurements of the $\sin^2\theta_{23}$ and Δm^2_{32} parameters from its muon neutrino disappearance analysis [7]. However a combine analysis of T2K's appearance and disappearance channels has become necessary in order to fully extract the four oscillation parameters that T2K is sensitive to: $\sin^2\theta_{23}$, Δm^2_{32}, $\sin^2\theta_{13}$, and δ_{CP}. Indeed with the increased precision of the appearance and disappearance analyses, the dependence on more than one mixing angle of the oscillation probabilities cannot be ignored anymore. More precisely looking at the leading terms for the electron neutrino appearance and muon neutrino disappearance:

$$P(\nu_\mu \to \nu_e) \approx \sin^2\theta_{23} \cdot \sin^2 2\theta_{13} \cdot \sin^2\left(\frac{\Delta m^2_{32} L}{4E_\nu}\right) \tag{1}$$

$$P(\nu_\mu \to \nu_x) \approx 1 - (\cos^4\theta_{13} \cdot \sin^2 2\theta_{23} + \sin^2\theta_{23} \cdot \sin^2 2\theta_{13}) \sin^2\left(\frac{\Delta m^2_{32} L}{4E_\nu}\right) \tag{2}$$

one sees that both θ_{13} and θ_{23} have a significant impact. Furthermore T2K is not currently capable of separating the contribution from θ_{13} and the CP violation δ_{CP}. The reactor neutrino measurement of $\sin^2 2\theta_{13}$ on the other hand is insensitive to δ_{CP} and can be therefore used to break the degeneracy between θ_{13} and δ_{CP} in the T2K data.

One of the approaches used by the T2K collaboration to combine its appearance and disappearance channels as well as the reactor neutrino experiment results on $\sin^2 2\theta_{13}$ was to perform a Bayesian inference analysis to construct posterior probability distributions for the four parameters and marginalize over all the systematic uncertainty parameters. In this approach the knowledge of the neutrino beam and neutrino cross sections as well as the reactor neutrino experiment results is included in the form of prior probability distributions. The ND280 detector measurement of the muon neutrino interaction rate prior to oscillation provides constraints on the neutrino beam and cross section parameters. The prior information, ND280 events and SK events are combined in a likelihood function to create the posterior distribution using a Markov Chain Monte Carlo since it cannot be computed analytically. The credible regions for δ_{CP} with respect to $\sin^2\theta_{13}$ show how the reactor neutrino results break the degeneracy between the two parameters (see Fig. 2(a)). The probability density for δ_{CP} after marginalization over all the other parameters shows a preference of the results for $\delta_{CP} \approx -0.5\pi$, however the credible interval at 90% extends to $\delta_{CP} = 0$ therefore this preference is not significant. Nonetheless these results demonstrate T2K's capacity to contribute to the neutrino community search for the value of δ_{CP}.

In May 2014 T2K has started collecting data with its beamline set in anti-neutrino beam mode. In this mode the ND280 detector, and in particular the charge measurement from its TPCs, will be essential to the oscillation measurement since the muon neutrino interaction rate in anti-neutrino beam mode is non negligible compared to the anti-muon neutrino and SK does not reconstruct the charge the charged leptons. The anti-muon neutrino disappearance and anti-electron neutrino appearance will provide crucial information for the δ_{CP} search and pave the way to a future discovery of this parameter by the next generation of long baseline neutrino experiments.

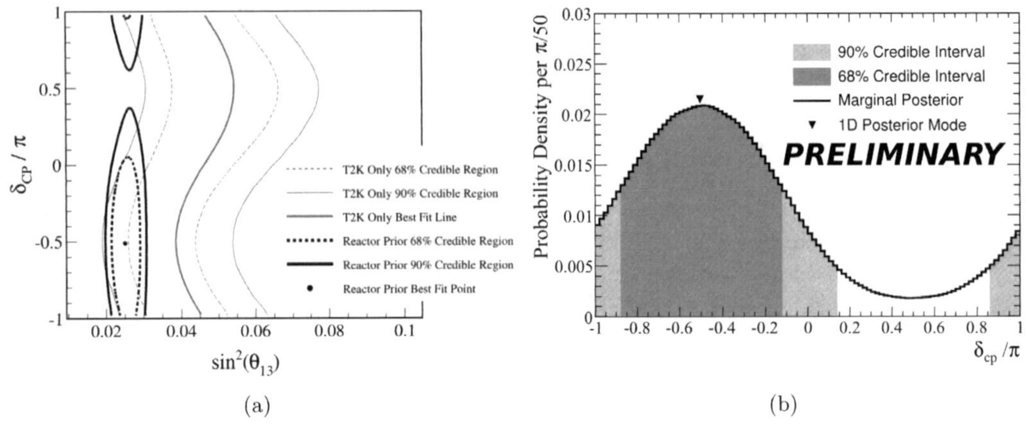

Figure 2. Credible intervals for the δCP parameters versus $\sin^2\theta_{13}$ and on the δCP probability density. The other oscillation and systematic uncertainty parameters are marginalized over.

4. Studies of the TPC electric and magnetic fields

The T2K TPCs play a central role in the ND280 analysis and therefore in the T2K oscillation results. For this reason the TPC calibration and reconstruction are always being developed and improved to try to reduce the systematic uncertainties relative to the TPC tracking. For example the reconstruction of vertical tracks was significantly improved in the latest version of the reconstruction software providing long vertical tracks from cosmic rays subsequently used to improve the gain calibration of the MicroMega detectors. This calibration is essential for the particle identification using the measured energy loss to identify muon tracks and select muon neutrino charge current interactions in the ND280 tracker.

Similarly a lot of work is dedicated to understanding the electric and magnetic fields in the TPCs. In ND280 the electric and magnetic fields are parallel to one another and perpendicular to the neutrino beam direction. The magnetic field was mapped prior to the installation of the detectors inside the ND280 magnet by a high precision mapping apparatus capable of reproducing the position of the magnetic probes with a precision of 10 μm. The resulting magnetic field map is use in the reconstruction of the data events to correct the displacement of the drift electrons due to the magnetic field distortions. A comparison of the momenta reconstructed in TPC2 and TPC3 from a single particle are used to verify the quality of the correction since TPC2 is near the center of ND280 where the field is very homogeneous while TPC3 is at the edge where significant field distortions are present. Applying the magnetic field correction brings the difference between the two reconstructed momenta from about 6% to less than 2% which is the target for the momentum bias.

The TPCs are also equipped of a laser system at 266 nm which illuminates aluminium dots located on the cathode foil in the middle of each TPC in order to create photo-electrons. The measured position of each dot is compared to the expected position in order to extract the integrated displacements undergone by the photo-electrons as they drifted from the cathode to the readout plane.

Finally the last tool at our disposition is looking at the residuals of a control sample of reconstructed tracks. The advantage of this technique is to be able to probe different parts of the TPC drift volume. Using this technique we have discovered the presence of electric field distortions of identical shape and slightly different amplitude in the six drift volumes. This simple empirical 2D model for the electric potential is capable of reproducing the overall features

Journal of Physics: Conference Series **650** (2015) 012014 doi:10.1088/1742-6596/650/1/012014

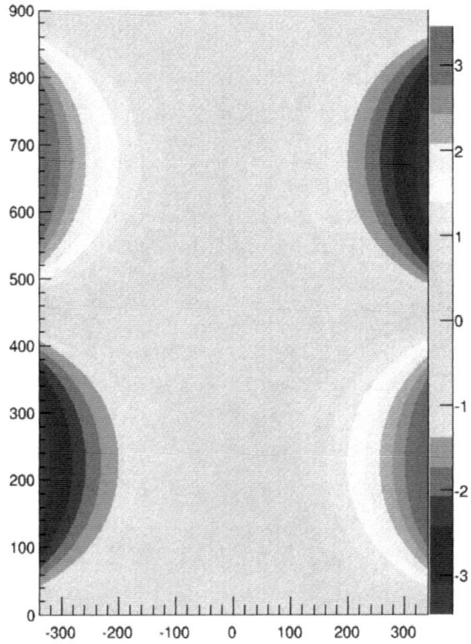

Figure 3. Z component(parallel to the neutrino beam direction) of the electric field distortions.

of the residuals observed in the data:

$$\varphi(x,z) = A\left[\sin\left(\frac{2\pi}{L}x\right)\cosh\left(\frac{z}{b_1}\right) + c\sin\left(\frac{6\pi}{L}x\right)\cosh\left(\frac{z}{b_2}\right)\right] \tag{3}$$

where x is along the drift direction, z is aligned to the neutrino beam, L is the maximum drift distance and A, b_1, b_2, and c are parameters tuned to the data. As shown on Fig. ??, the distortion displays a symmetric structure. The symmetry along the x axis means that electrons drifting from the cathode are displaced first towards positive (negative) y and then towards negative (positive) y but the same amount. As a consequence the integrated displacement observed using the photo-electrons from the laser system is insensitive to these electric field distortions. We are currently improving the empirical model and in parallel we are investigating the possible sources of this electric field distortion using the modeling software COMSOL in order to simulate the TPC electric field. Various deformations of the TPCs such as a bulging of the TPCs walls were tested however the corresponding electric field distortion cannot reproduce our observations in the data.

5. TPC gas interaction analysis

Besides its successes with oscillation measurements, T2K has also provided the neutrino community with multiple cross section measurements. Most of these results were produced using the near detectors and in particular the ND280 tracker for results such as the first electron neutrino cross section measurement at a few GeV energy since the Gargamelle measurement in 1976 [8]. Even a neutral-current interaction cross section on oxygen was performed by observing nuclear deexcitation γ rays in SK [9].

Of all the possible cross section measurement in T2K, the most unique is the possibility to use the T2K TPCs to measure neutrino-argon cross section interaction in a gas phase. Indeed the

T2K TPCs are currently the only gaseous TPCs exposed to an accelerator neutrino. Although the low interaction rate with the TPC gas at 1 atm will lead to statistically limited measurements, any results will be very useful to the neutrino community all the charged particles exiting the nucleus after the neutrino interactions should be detectable, even protons of very low energy. These protons are particularly useful to understand the nuclear effects involved in the neutrino interactions on heavy nuclei.

As a neutrino interacts with a nucleus, it does not always interact with a single neutron to produce a lepton and a proton. Some interactions involve neutrons in a bound state inside the nucleus leading to the emission of additional nucleons and/or pions. Hadronic models are being developed to describe these interactions because they modify the overall neutrino cross section on the nucleus significantly. For example these nuclear effects can explain the apparent discrepancy in the quasi-elastic cross section measurements on carbon from the MiniBooNE and NOMAD experiments [10]. The final state interactions (FSI) on the nucleons complicate the study of the hadronic effects by scattering or absorbing the pions and protons as they exit the nucleus.

Understanding these nuclear effects is very important T2K but also for future long baseline neutrino experiments such as LBNE or LBNO. Both experiments plan to use liquid argon TPCs as their far detector which is why many liquid argon TPC experiments are in preparation or operating already as R&D projects. Some of these experiments, such as ArgoNeuT and MicroBooNE have performed or will perform neutrino-argon cross section measurements to help develop neutrino interaction models. A cross section result from the T2K TPCs would be very complementary to the liquid argon effort because the T2K TPC gas is composed at 95% of argon and it is at a neutrino energy regime where charged current quasi elastic interactions dominate. The ArgoNeuT experiment reported a minimum detection threshold for protons at about 200 MeV/c [11] while we expect to reach 30 MeV/c with the T2K TPCs. The interval below 200 MeV/c covered by the T2K TPCs is particularly interesting because it is expected to be dominated by protons having undergone FSI. Furthermore this interval also displays large discrepancies in the predicted momentum distributions from the NEUT [12] and GENIE [13] neutrino interaction generators as shown on Fig. 3.

Currently the biggest task to develop the neutrino-gas interaction analysis in T2K TPC is the implementation of a new TPC reconstruction. Indeed although a reconstruction algorithm exists and has been used to produce the published T2K results, it was designed to reconstruct tracks from particles originating from outside the TPCs. This is why a new reconstruction code called TREx for TPC reconstruction extension was written and is currently undergoing validation. TREx was designed with the concept of vertex connecting multiple tracks and can therefore identify and handle the gas interaction vertices inside the TPCs. Components from the previous TPC reconstruction were reused such as the maximum likelihood fit of the tracks providing a bias and a resolution respectively below 2% and 10% at 1 GeV/c. The challenge of the reconstruction and the analysis will be to discern the signal from the large background of tracks originating from the detectors surrounding the TPCs and in particular from the lead of the ECals. The first neutrino-gas interaction results are expected in 2015.

6. Conclusion

By performing a combined analysis of its neutrino electron appearance and neutrino muon disappearance data, T2K has demonstrated its capacity to provide in the future, given more data, the first hints of the CP phase. To guarantee that future measurements are not systematics limited, ongoing work is dedicated to understanding the source of the TPCs electric field distortions that were discovered recently. Improvements to the TPC reconstruction are also underway to allow the reconstruction of neutrino interactions happening inside the TPC themselves and perform very promising measurements of these interactions and their byproducts.

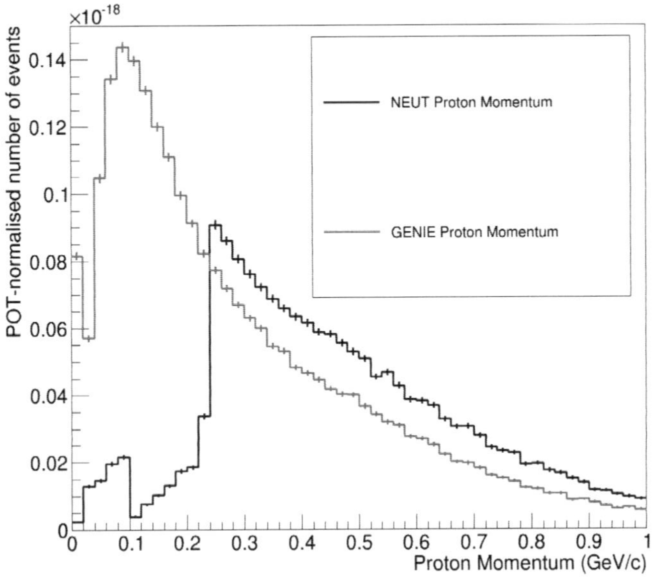

Figure 4. Neutrino generator prediction for the momentum distribution of all the protons emitted from neutrino interactions on argon.

References

[1] Abe K. *et al.* 2011 *Nucl. Instrum. Methods* A **659** 106
[2] Abgrall N. *et al.* 2011 *Nucl. Instrum. Methods* A **637** 25
[3] Amaudruz P.-A. *et al.* 2012 *Nucl. Instrum. Methods* A **696** 1
[4] Assylbekov S. *et al.* 2012 *Nucl. Instrum. Methods* A **686** 1
[5] Allan D. *et al.* 2013 *JINST* **8** P10019
[6] Abe K. *et al.* 2014 *Phys. Rev. Lett.* **112** 061802
[7] Abe K. *et al.* 2014 *Phys. Rev. Lett.* **112** 181801
[8] Abe K. *et al.* 2014 *Phys. Rev. Lett.* **113** 241803
[9] Abe K. *et al.* 2014 *Phys. Rev.* D **90** 072012
[10] Formaggion J. A. and Zeller G. P. 2012 *Rev. Mod. Phys.* **84** 1307
[11] Acciarri R. *et al.* 2014 *Phys. Rev.* D **90** 012008
[12] Hayato Y 2009 *Acta Phys. Pol.* B **40** 2477
[13] Andreopoulos C *et al.* 2010 *Nucl. Instrum. Methods* A **614** 87

Journal of Physics: Conference Series **650** (2015) 012015 doi:10.1088/1742-6596/650/1/012015

The WA104 Experiment at CERN

Maurizio Bonesini
(on behalf of the ICARUS/WA104 Collaboration)
Sezione INFN Milano Bicocca,
Dipartimento di Fisica G. Occhialini,
Piazza Scienza 3, 2016 Milano
Italy

E-mail: maurizio.bonesini@mib.infn.it

Abstract. The refurbishment operations, inside the WA104 programme at CERN, of the Icarus T600 detector, in view of its possible run on the FNAL short-baseline Booster neutrino beam (BNB), will be outlined.

1. Introduction

In recent years, in disagreement with the established picture of three neutrinos [1], several experimental "anomalies" both at reactors and accelerators have been reported [2]. They hint to new physics with the possibile existence of one or more "sterile" neutrinos with masses at or below the few eV range. An important contibution to the sterile neutrino search has been recently obtained by the ICARUS Collaboration at LNGS on the CNGS beam, as also reported at this conference [3].

New searches for sterile neutrinos have been proposed a reactors, looking for oscillations with $L/E \sim 1m/MeV$ [4] or using intense radioactive sources [5]. In addition, the nuSTORM project, based on a low energy neutrino factory, for a precision search of sterile neutrinos has been proposed [6]. With regards to the present scenario, the best opportunity will be provided however by a dedicated accelerator-based neutrino experiment, where the presence of the oscillation signal in ν_e appearance and disappearance modes as well the ν_μ disappearance can be investigated with the same experimental setup.

A new experiment with a set of three liquid Argon (LAr) TPC detectors (LAr1-ND, MicroBooNE and ICARUS T600), at different baselines along the Boster Neutrino Beam (BNB) at FNAL, has been recently proposed [7] and will soon allow a very sensitive search for $\Delta m^2 \simeq 1eV^2$ neutrino oscillations in different channels. The ICARUS T600 detector will operate as far detector at 600 m and the WA104 programme at CERN has therefore been conceived for the needed refurbishment operations.

While, in the deep underground conditions of the LNGS laboratory, a single prompt trigger has always ensured the unique timing connection of the main image of the event for the T600 detector, at FNAL, at shallow depths, several additional cosmic muons (~ 10) will be present in the 1 ms drift time, giving problems for track reconstruction. A 4π, segmented cosmic muon tagging system (CTS) to identify cosmic rays entering the detector, will have to be designed and constructed. Available technologies to realize this system (area ~ 1000 m^2) include either resistive plane chambers (RPC) or scintillator slabs read out by photomultipliers or SiPM.

In addition, the system for the detection of the light emitted in LAr (at $\sim$ 128 nm) by ionizing particles will have to be upgraded, to allow both a more precise event timing and localization and the exploitation of the bunched beam structure of the BNB beam at FNAL (1.15 ns every 19 ns) to reject out-of-bunch cosmics, as proposed in [8].

2. The ICARUS T600 detector

The present ICARUS T600 cryogenic detector is made of two identical modules, $3.6 \times 3.9 \times 19.6\text{m}^3$, for a total mass of $\sim$ 760 tons of Liquid Argon ($\sim$ 476 active mass). Each module is equipped with 2 chambers on the long sides, with three readout planes with wires at $0^0, \pm 60^0$. With a 3 mm pitch and a 3 mm plane spacing, there are about 54000 readout wires. The two TPCs of each module share a common cathode, placed in the middle. Electrons have a maximum drift length of 1.5 m, in a uniform electric field of 500 V/cm. A schematic layout of the detector is shown in Figure 1 (for more details see also [9]).

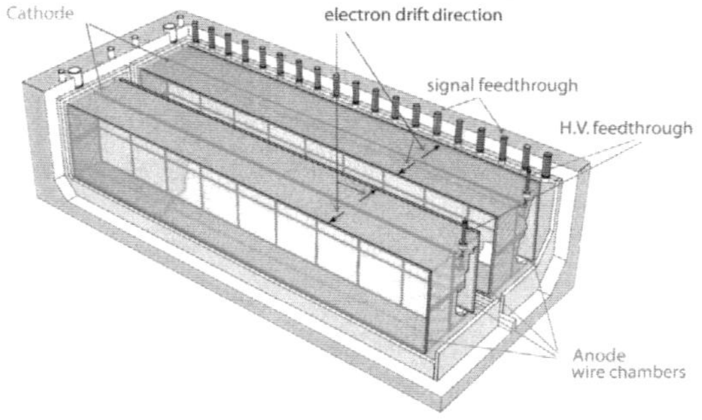

Figure 1. Schematic layout of the present T600 detector.

The LAr TPCs are presently complemented by a system of 54+28 8" PMTs, coated with a TPB wavelength shifter, to detect the scintillation light emitted at $\sim$ 128 nm by crossing ionizing particles [10].

The T600 detector has been exposed to the CNGS neutrino beam in 2010-2012, with a total statistics of 8.6×10^{19} p.o.t and a detector livetime in eccess of 93%.

A key feature of the T600, in view of new larger mass LAr TPCs, is the achievement of very long electron mobility [11]. To ensure long drift path for ionisation electrons without appreciable attenuation, the level of electronegative impurities in LAr must be kept exceptionally low (order some ppt). A lifetime in excess of 16 ms ($<$ 20 ppt O_2 equivalent) was obtained since April 2013, thus demonstrating the effectiveness of single-phase LAr-TPC technique for huge detectors (up to 5m drift).

During its operation, the T600 detector demonstrated excellent detection properties, such as:

- good 3D reconstruction capabilities ($\sim 1\ mm^3$ spatial resolution), with accurate ionization measurement;
- good calorimetric energy reconstruction ($\sigma(E)/E \simeq 0.03(0.30)/\sqrt{E}(GeV)$ for e.m. (hadronic) showers;
- good particle identification via dE/dx measurement;
- momentum reconstruction of non contained μ via Multiple Scattering ($\Delta p/p \simeq 16\%$ in the 0.4-4 GeV/c range of interest for the future short and long baseline ν experiments).

3. Foreseen upgrades of the T600 detector: the WA104 programme

The T600 detector has been transported to CERN at the end of 2014 for overhauling and refurbishment, in view of its future use on the FNAL BNB beam as far detector [7] or on the new LBNF long-baseline beam, as a possible near detector. One module, inside the clean room at CERN Bldg 185, is shown in figure 2.

Figure 2. One module of the T600 detector inside the clean room at CERN Bldg. 185.

The WA104 programme at CERN foresees the following main operations on the T600 detector:

- use of new cold vessels and new purely passive thermal insulation, refurbishing in parallel the cryogenic and purification equipment;
- implement a new cathode with better planarity;
- upgrade of the light collection system with new PMTs;
- upgrade of the existing "warm" electronics.

Futher *R&D* activities will be carried on in parallel (including some specifically related to the future LBNF programme at FNAL) and will encompass:

- design of a cosmic tagging system for operations at shallow depths of the LAr TPCs;
- study of magnetization for the LAr TPCs;
- consequent replacement of the PMTs with photodetectors (such as the SiPM) insensitive to external magnetic fields;
- LAr doping.

3.1. New thermal insulation and vessels

A purely passive insulation has been chosen for the refurbishing of the T600 at CERN, coupled to a standard cooling shield with boiling N_2. The technique has been developed since 50 years and is widely used for large storage vessels and ships for liquefied natural gas [12]. Polyurethane foam has been chosen as insulation, with a thickness of 600 mm for the bottom and lateral sides and 400 mm for the top side. With this configuration, the heat loss through the insulation will

Journal of Physics: Conference Series **650** (2015) 012015 doi:10.1088/1742-6596/650/1/012015

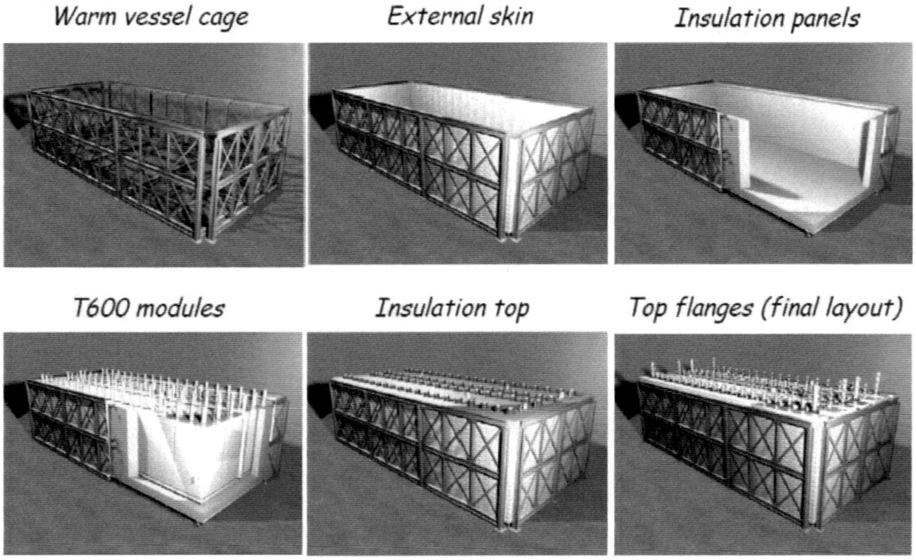

Figure 3. New T600 layout, including warm vessel and insulation panels.

be $\sim$ 6.6 kW, giving a total heat load of $\sim$ 12 kW, including all external heat contributions (cables, pumps, ..) The adopted layout is shown in Figure 3.

Taking into account the expected heat loss, about 100K (273K) liters of LN2 (LAr) will be needed in the cooling phase of the detector for each module of the T600. The original scheme of the ICARUS T600 cryogenics and LAr purification systems will be preserved, re-using most of the present plant.

3.2. Upgraded light collection system

Operations of the T600 detector at shallow depths will require an improved light collection system, able to work with energy depositions down to $\sim$ 100 MeV. The new light collection system will have to localize the track associated with the light pulse with an accuracy better than 1 m along the longitudinal z coordinate, to allow efficient cosmic muons rejection. With the PMTs' layout shown in figure 4, a reconstruction in the z coordinate with a FWHM of 19, 24 or 38 cm will be obtained with 27, 54 or 90 PMTs per module side. To exploit the BNB bunch structure, for futher rejection of out-of-bunch cosmic events, an overall time resolution $\sim$ 1 ns will be needed. This will require an accurate time calibration of the PMTs system.

Tests are under way to select the most suitable PMT model [13]. Three large area (8") PMTs: Hamamatsu R5912 Mod and R5912-02 Mod and ETL 9357 KFLB are under study, both at room and cryogenic temperature. Their main characteristics are shown in Table 1.

Table 1. Main characteristics of used 8" PMTs.

	Hama R5912	Hama R5912-02	ETL 9357KFLB
no. dynodes	10	14	12
Gain (typ)	10^7	10^9	10^7
	(at 1500 V)	(at 1700 V)	(at 1500 V)
risetime (ns)	3.8	4	3.5
TTS (FWHM ns)	2.4	2.8	4
dark current(nA)	50	10^3	10
Q.E.at 390 nm (%)	25	25	18

These PMTs have borosilicate glass windows and bialkali photocathode with platinum undercoating, to restore photocathode conductivity at low temperatures. Sensitivity in the

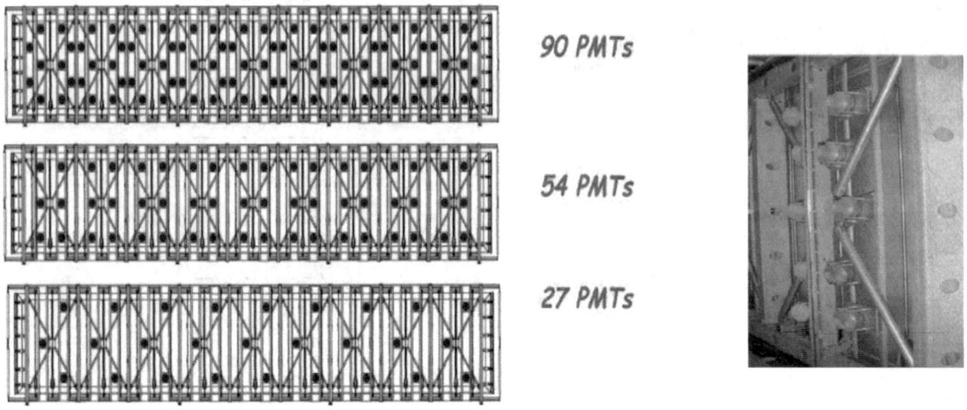

Figure 4. Left panel: foreseen PMTs layout for one side of one module. Right panel: rendering of the installation of PMTs inside one module.

VUV range is obtained via a coating of the sanded photocathode window with a WLS (TPB in this case). The quantum efficiency (Q.E.) in the VUV region was measured with a McPherson 234/302 VUV monochromator, equipped with a Mc Pherson 632 Deuterium lamp. The right panel of figure 5 shows the Q.E. measured for these PMTs. For cryogenic tests, PMTs are directly immersed in a LN2 bath (T=77 K) and measurements are taken after few days of rest. Gain both at room and cryogenic temperatures are shown in the left panel of figure 5. All PMts show good photocathode uniformity and a linear behaviour.

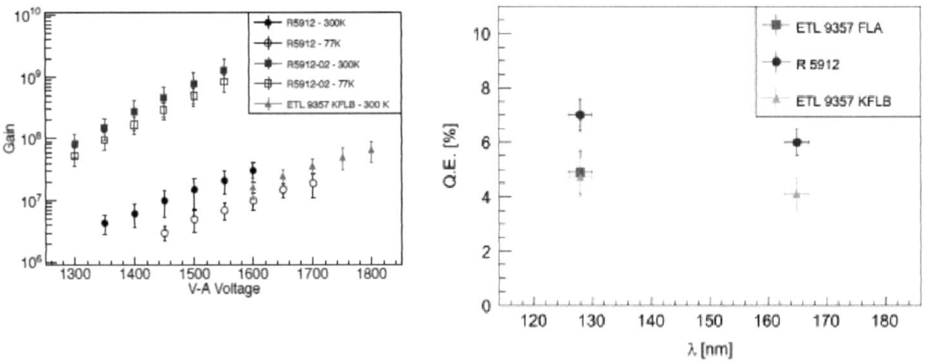

Figure 5. Left panel: gain as a function of HV, for the different types of tested PMTs. Right panel: Q.E. for the different types of tested PMTs, coated with TPB.

For gain, linearity and photocathode uniformity measurements, PMTs were illuminated with a 405 nm NICHIA NDV1413 laser diode, using an Avtech AVO-9A-C-P2-LARB pulse generator

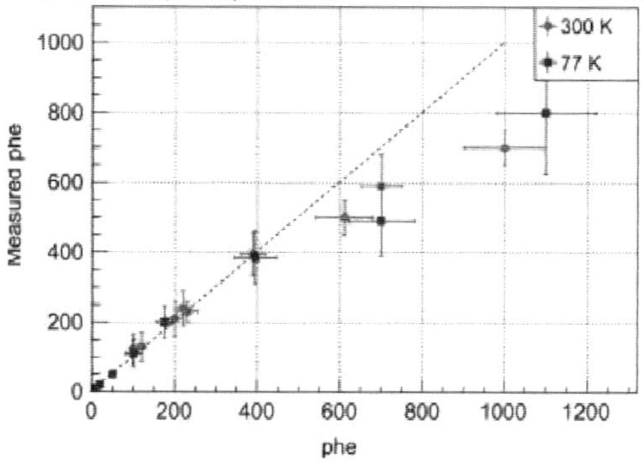

Figure 6. Linearity for the Hamamatsu R5912 PMT under test, at room and cryogenic temperatures.

and a connecting 7μm core, 3m long optical fiber. The linearity of one Hamamatsu R5912 PMT, both at room and cryogenic temperatures, is shown in figure 6.

3.3. New electronics

The architecture of the T600 wires electronics was based on a custom design analogue low noise amplifier, a multiplexed 2.5 MHz 10-bit ADC and the VME backplane for data transfer. A S/N ratio better than 10 and a single point spatial resolution around 0.7 mm were thus obtained. The new design of the electronics chain foresees:

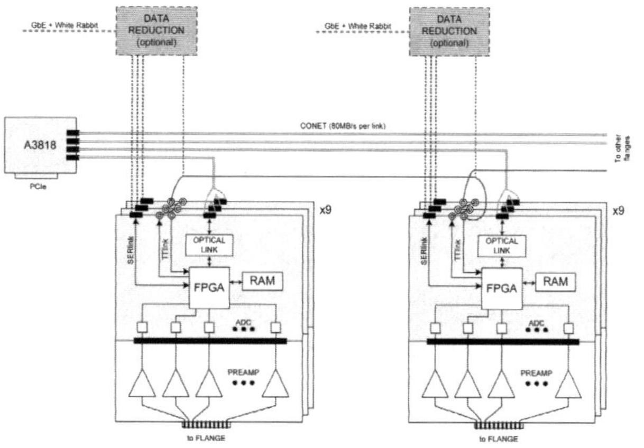

Figure 7. Schematic layout of the prototype implementing the readout system, based on a modern switched I/O on a low cost optical Gigabits serial link.

- a modern serial switched I/O for data flow, giving transfer rates up to a few Gb/s (see Figure 7 for details), as compared to the 8-10 Mb/s of the VME standard;
- one serial (10-12 bits) ADC per channel, allowing synchronous sampling within the 400 ns sampling window;
- a more compact design for both analogue and digital electronics (512 channels) integrated in the ad-hoc flange design (see Figure 8);
- the digital part is contained in a single FPGA per board, that handles the signal filtering and organizes the infos provided by the ADCs.

A cold front-end option is also under study and prototypes, developed in collaboration with BNL, will be tested at CERN.

Figure 8. Left panel: prototype of the feed-through flange under test. Right panel: first prototypes of new boards under test at LNL

4. Conclusions

The T600 refurbishment plans, inside the WA104 experiment, have been outlined and the detector is expected to be ready in time for the start of operations at FNAL BNB neutrino beam, in the near future. Aim of the experiment at FNAL will be a definitive clarification of the "LSND anomaly" and the collection of a significant amount of data in the energy range of interest for future long-baseline experiments.

References

[1] K. Olive *et al.* (PDG), Chin. Phys. **C38** (20140 09001.
[2] G. Mention *et al.*, "The Reactor Antineutrino Anomaly", Phys. ReV. **D83** (2011) 073006 ;
W. Hampel *et al.* (GALLEX Coll.), "Final Results of the Cr-51 neutrino source experiments in GALLEX", Phys. Lett. **B420** (1998) 114;
J. Abdurashitov *et al.* (SAGE Coll.), " Measurement of the response of the Russian-American Gallium experiment to neutrinos from a Cr-51 source", Phys. ReV. **C59** (1999) 2246 ;
A. Aguilar-Arevalo *et al.* (LSND Coll.), "Evidence for neutrino oscillations from the observation of $\overline{\nu_e}$ appearance in a ν_μ beam", Phys. ReV. **D64** (2001) 112007;
A. Aguilar-Arevalo *et al.* (MiniBooNE Coll.), "Improved Search for $\overline{\nu_\mu} \mapsto \overline{\nu_e}$ oscillations in the MiniBooNE experiment", Phys. ReV. Lett. **110** (2013) 161801.
[3] J. Kisiel *these proceedings*;
M. Antonello *et al.*, "Experimental search for the "LSND anomaly" with the ICARUS detector at the CNGS neutrino beam", Eur. Phys. J. **C73** (2013) 2599.
[4] A.S. CuCoanes *et al.* (Nucifer Coll), "Status of the Nucifer experiment", J. Phys. Conf. Ser. **375** (2012) 042063.
[5] G. Bellini *et al.* (Borexino Coll.), "SOX: Short distance neutrino Oscillations with Borexino", JHEP 1308 (2013) 038
[6] P.Kyberd *et al.*, "nuSTORM: Neutrinos from Stored Muons", arXiv:1206.0294 (2012) ;
D. Adey *et al.*, "Light sterile neutrino sensitivity at the nuSTORM facility", Phys. ReV. **D89** (2014) 071301.
[7] M. Antonello *et al.*, " A proposal for a Three Detector Short-baseline ν oscillation Program in the Fermilab Booster Neutrino Beam ",LOI, FNAL, 2015.
[8] C. Rubbia "Using the beam bunch-structure with the T600 detector", SBN internal report, 2014.
[9] J. Kisiel *these proceedings*;
S.Amerio *et al.*, "Design, construction and tests of the Icarus T600 detector", Nucl. Instr. and Meth. **A527** (2004) 329;
C. Rubbia *et al.*, "Underground operations of the Icarus T600 LAr-TPC, first results", JINST 6 (2011) P07011.
[10] P. Cennini *et al.*, "Detection of scintillation light in coincidence with ionizing tracks in a liquid argon time

projection chamber", Nucl. Instr. and Meth. **A432** (1999) 240;
P. Benetti *et al.*, "Detection of the VUV liquid argon scintillation light by means of glass-window photomultiplier tubes", Nucl. Instr. and Meth. **A535** (2003) 89;
A. Ankowsky *et al.*, " Characterization of ETL 9357FLA photomultiplier tubes for cryogenic temperature operations", Nucl. Instr. and Meth. **A556** (2006) 149.

[11] M. Antonello *et al.*,"Experimental observation of an extremely high electron lifetime with the ICARUS-T600 LAr-TPC", JINST 9 (2014) P12006.

[12] http://www.gtt.fr

[13] A. Falcone *et al.*, " Comparison between large area photomultiplier tubes at cryogenic temperature for neutrino and rare event physics experiments", Nucl. Instr. and Meth. **A** (2014), in press.

Journal of Physics: Conference Series **650** (2015) 012016 doi:10.1088/1742-6596/650/1/012016

HARPO: beam characterization of a TPC for gamma-ray polarimetry and high angular-resolution astronomy in the MeV-GeV range

Shaobo Wang
LLR, Ecole Polytechnique, CNRS/IN2P3, France
E-mail: wang@llr.in2p3.fr

Denis Bernard, Philippe Bruel, Mickael Frotin, Yannick Geerebaert, Berrie Giebels, Philippe Gros, Deirdre Horan, Marc Louzir, Patrick Poilleux, Igor Semeniouk
LLR, Ecole Polytechnique, CNRS/IN2P3, France

David Attié, Denis Calvet, Paul Colas, Alain Delbart, Patrick Sizun
Irfu, CEA-Saclay, France

Diego Götz
AIM, CEA/DSM-CNRS-Université Paris Diderot, France
Irfu/Service d'Astrophysique, CEA-Saclay, France

Sho Amano, Takuya Kotaka, Satoshi Hashimoto, Yasuhito Minamiyama, Akinori Takemoto, Masashi Yamaguchi, Shuji Miyamoto
LASTI, University of Hyôgo, Japan

Schin Daté, Haruo Ohkuma
JASRI/SPring8, Japan

Abstract. A time projection chamber (TPC) can be used to measure the polarization of gamma rays with excellent angular precision and sensitivity in the MeV-GeV energy range through the conversion of photons to e^+e^- pairs.

The Hermetic ARgon POlarimeter (HARPO) prototype was built to demonstrate this concept. It was recently tested in the polarized photon beam at the NewSUBARU facility in Japan. We present this data-taking run, which demonstrated the excellent performance of the HARPO TPC.

1. Introduction

Gamma-ray astronomy is the primary means by which we can study the non-thermal processes occuring in cosmic sources such as active galactic nuclei (AGN), pulsars and gamma-ray bursts (GRBs) [1]. The ability to measure the linear polarization in this energy range would provide a powerful diagnostic for the understanding of the physical processes at work in these sources [2]. A telescope with the ability to perform polarimetry above 1 MeV has never been flown in space [3]. Gamma-ray telescopes have improved in sensitivity and resolution from COS-B [4] to the Fermi-LAT [5], but the performance of these pair-creation telescopes decreases at low energy where the background is problematic and the angular resolution is limited by multiple scattering.

The performance and polarimetry potential of a TPC have been studied in detail in Ref. [6] and [7]. Several effects contribute to the angular resolution, which is shown as a function of photon energy in Fig. 1 for an argon gas TPC. The angular resolution for pair production is limited by multiple scattering of the electrons in the gas. Also, most conversions take place in the field of a nucleus whose recoil momentum is impossible to measure because of the very short recoil path length. This puts a limit on the resolution at energies $< 100\,\mathrm{MeV}$. Even with these limitations however, an improvement in angular momentum of up to an order of magnitude with respect to that of the Fermi-LAT is achievable with a TPC.

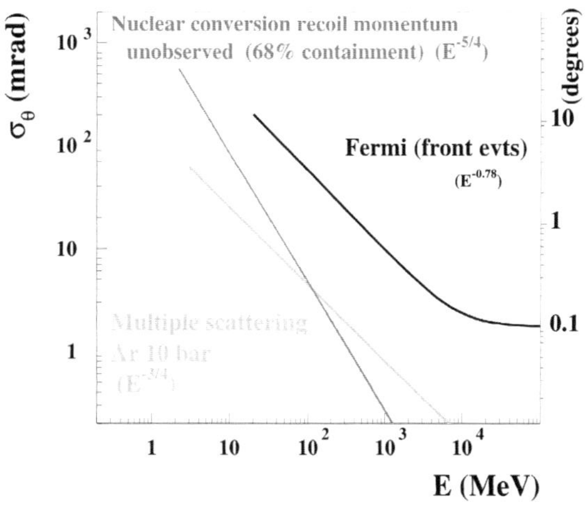

Figure 1. Various contributions to the angular resolution of a TPC gamma telescope as a function of the photon energy [6], compared to that of the Fermi-LAT [5]. The multiple scattering contribution (green) is given for a 10 bar argon TPC, using an optimal tracking such as that implemented in a Kalman filter, and with a track sampling of 1 mm and a spatial resolution of 0.1 mm. At high energies ($> 100\,\mathrm{MeV}$), it is limited by the multiple scattering of the electrons in the gas. At lower energies, it is constrained by the unknown recoil momentum of the nucleus. This limit can be overcome by looking at triplet conversion, but they are rarer and more difficult to reconstruct.

The polarimetry of cosmic sources with a pair-creation telescope has long been considered to be difficult or even impossible, as thick, high-Z detector elements are needed to convert the incoming photon, and the conversion electrons then undergo multiple scattering in the converter, after which the information about the azimuthal angle of the conversion plane is blurred [8]. We have built and validated an event generator [7] which is full (5D, either nuclear or triplet conversion) exact down to threshold and polarized. We have characterized the properties of a telescope based on a thin homogeneous detector with optimal tracking and have established the power laws that describe the dominating contributions to the angular resolution [6]. For such a detector, even when the dilution of the effective polarization asymmetry due to multiple scattering is taken into account, polarimetry can still be performed with high precision (still under the assumption of optimal tracking) [7].

The HARPO project aims to characterize the TPC technology as a high angular resolution polarimeter and telescope in the MeV-GeV range, enabling us to obtain unprecedented sensitivity for the detection of low-energy gamma rays. A TPC is a detector in which traversing charged particles ionize the detector material. The electrons produced drift along an electric field, E, and are then amplified and measured on the x-y readout plane. The drift time gives a measure of the z coordinate. HARPO would be the first space polarimeter above 1 MeV.

In this paper we first describe the demonstrator that we have built to validate the performance of the TPC technology, in particular the characterization of the GEM and micromegas combinations used for gas amplification [1]. Then we present the recent experimental campaign, in which the detector was exposed to the quasi-monochromatic and almost fully polarized beam provided by the BL01 line of the NewSUBARU facility, operated by the LASTI in University of Hyôgo in Japan.

2. The HARPO TPC demonstrator

The HARPO demonstrator [9] (Fig. 2) is a 30 cm cubic TPC which can be operated from low pressure up to 5 bar. It is surrounded by 6 scintillator plates that provide an external trigger. Amplification is performed with the combination of two GEMs [10] and one micromegas [11].

The readout comprises two series (x, y, perpendicular to each other) of 288 copper strips with 1 mm pitch. Signals are acquired with AFTER chips [12] using the Feminos system [13].

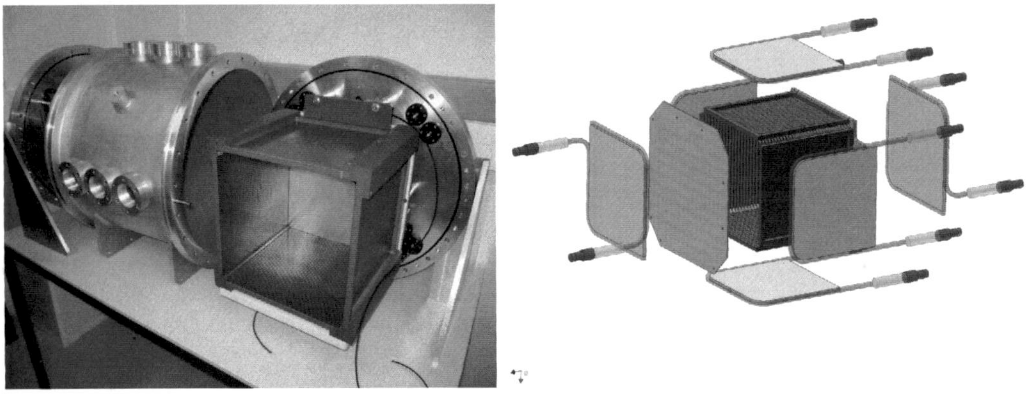

Figure 2. Left: the HARPO TPC demonstrator. Right: sketch of HARPO. A cubic TPC in the center, with a readout plane of micromegas and GEM on the left. Six scintillator plates, each equipped with a wavelength shifter and two photo-multipliers, are used for trigger. The system fits in an aluminum cylinder which can operate up to 5 bar pressure.

[1] The 2012 tests showed that with the 0.4 mm narrow collecting strips that we are using, the micromegas alone does not provide sufficient amplification at 4 bar for routine operation in a safe configuration [9]. We have therefore complemented the micromegas with two layers of Gas Electron Multiplier (GEM).

Journal of Physics: Conference Series **650** (2015) 012016 doi:10.1088/1742-6596/650/1/012016

3. The HARPO amplification system

We first characterized the performance of the GEM+micromegas amplification system using a radioactive source. The amplification system was subsequently integrated into the TPC detector, where it was characterized using cosmic rays.

3.1. Characterization of micromegas and GEM(s) combinations with a radioactive source

We characterized the combination of a micromegas and either one or two GEM in a gas mixture (Ar:95 %Isobutane:5 %) at atmospheric pressure. This was done in a dedicated test setup using a ^{55}Fe source[16]. The successive amplification steps were kept at a distance of 2 mm from each other by spacers (Fig. 3 left).

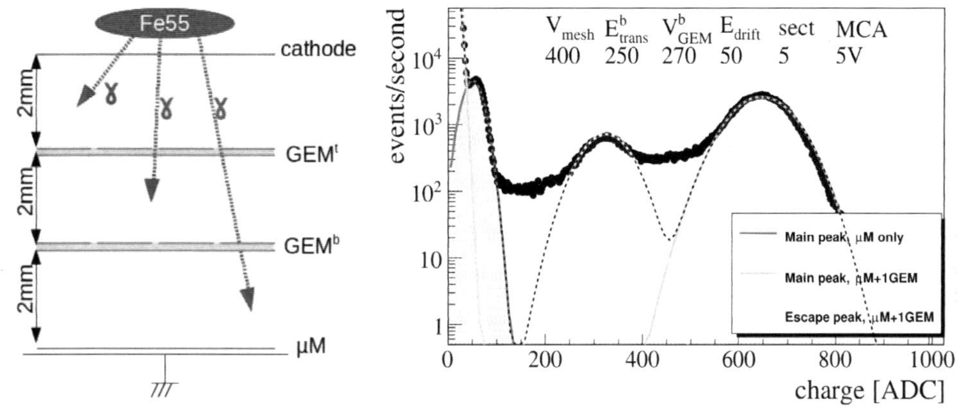

Figure 3. Left: the layout of the test setup, with one micromegas and two GEMs. (By applying a null or inverted voltage to the top GEM, we only observe conversion in the two lower regions). Right: an example of a measured ^{55}Fe spectrum from amplification with one micromegas and one GEM. The main peak (5.9 keV) is plotted after amplification with only the micromegas (red) or with both micromegas and GEM (green). The escape peak is only visible with full amplification (yellow).

In argon, the X-rays from a ^{55}Fe source deposit either 5.9 keV (main peak) or 2.7 keV (escape peak) upon ionization. This conversion can take place either above or below a given GEM sheet. The ionization electrons are therefore either amplified by that GEM or not. A typical spectrum is shown in Fig. 3 (right). The main and the escape peaks are seen with amplification from the micromegas and one GEM, and the main peak with micromegas amplification only. The ratio of the position of the two main peaks provides a precise measurement of the absolute GEM amplification gain. Further details on these measurements, including foil transparency and extraction efficiency, can be found in Ref. [16].

3.2. Characterization with cosmic rays

The system of micromegas and 2 GEMs was commissioned in the detector. It was tested with cosmic rays, using the same gas mixture at several pressures. Most cosmic muons are relativistic and therefore deposit the same average energy per unit length. The TPC was set "vertical", so that most cosmic rays entered it through the amplification system. We selected events with a cosmic ray that crossed the full z of 30 cm by triggering on the coincidence of the up and down scintillator signals.

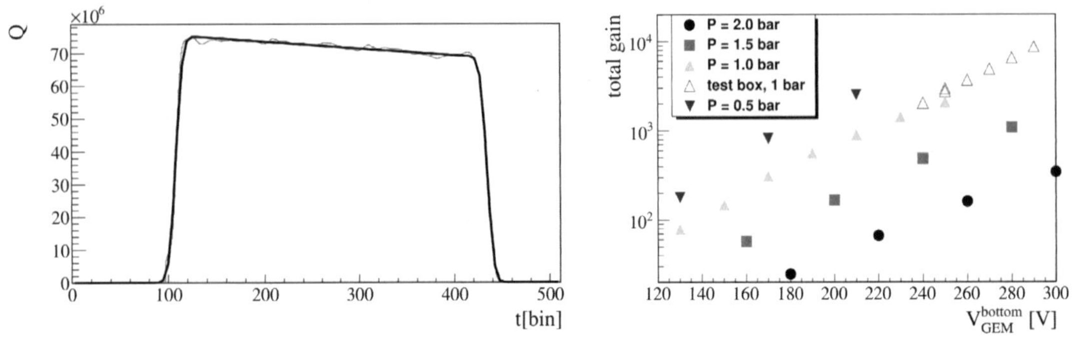

Figure 4. Left: average total charge per track as a function of time, for a run of cosmic rays that traversed the full TPC length, entering through the anode and exiting through the anode. We extract the drift velocity, the electron absorption per unit length and the amplification gain from such spectra. Right: the effective gain measured from cosmic rays in the TPC for several gas pressures. The results at 1 bar are consistent with those obtained with the radioactive source.

Table 1. Electron beam energy, laser wavelength, and γ-ray energy at the Compton edge obtained from these.

E_e^- [GeV]		0.618	0.982	1.233	1.480	Pulsing rate	Polarization
Laser	λ [µm]		E_γ [MeV]			[kHz]	
Nd:YVO_4 (2ω)	0.532	13.4	33.3	52.1	74.3	20	$P \approx 1$
Nd:YVO_4 (1ω)	1.064	6.76	16.9	26.6	38.1	20	$P \approx 1$
Er(fiber)	1.540	4.68	11.8	18.5	26.5	200	$P = 0$
CO_2	10.55		1.74	2.73	3.93	CW	$P = 0$

The charge distribution over drift time is shown in Fig. 4 (left). The value of the drift velocity, v_{drift}, was easily obtained using the basic relation of the TPC mechanism, $z = v_{\text{drift}} \times t_{\text{drift}}$, since the cosmic rays traverse the full z thickness of $L_{TPC} = 30\,\text{cm}$. The slope of the plateau corresponds to electron absorption in the gas. The average charge per track for each run is used to estimate the amplification gain. From this spectrum, the drift velocity, the attenuation length and the total amplification gain were extracted. Fig. 4 (right) shows the amplification gain for several TPC gas pressures as a function of the voltage on one of the GEMs. The measurements are consistent with those obtained using the ^{55}Fe source.

4. Data-taking at NewSUBARU

In Nov. 2014 the detector was exposed to a beam of pseudo-monochromatic gamma-ray photons delivered by the BL01 beam line at NewSUBARU [15].

4.1. Laser-Compton Source (LCS)

The gamma-ray beam is produced by the inverse Compton scattering of laser photons by relativistic electrons. The electron beam energy can be varied in the range 0.6 $\sim$ 1.5 GeV. The laser beams available are: Nd:YVO_4 (2ω) with wavelength $\lambda = 0.532$ µm, Nd:YVO_4 (1ω) $\lambda = 1.064$ µm, Er (fiber) $\lambda = 1.540$ µm and CO_2 $\lambda = 10.55$ µm. This results in a gamma energy range at the Compton edge between 1.7 MeV and 74 MeV as shown in Table 1. As the gamma has a maximum energy for forward Compton scattering, energy selection was performed by using a collimator on axis. When collimation is applied, the polarisation of the laser beam is almost entirely transferred to the gamma beam: in that way, an almost fully polarized gamma beam ($P \approx 1$) is obtained. A general issue in polarization studies is the control of a possible systematic bias induced by the non-cylindrical-symmetric structure of the detector [17]. Therefore in addition to the fully polarized data, some data with a non polarized beam were taken. To produce such a beam, a $\lambda/4$ plate is used. This changes the linear polarization to circular polarization, which, as far as pair conversion is concerned, is equivalent to random polarization ($P = 0$). Or vice versa.

4.2. The HARPO Trigger system

For the previous characterizations using cosmic rays, the TPC was triggered on a simple coincidence of scintillator signals [9, 14]. In contrast, for data taking in beam, we wished to maximize the fraction of selected gammas that converted in the gas. Therefore a dedicated, more sophisticated trigger was designed. In addition it provides separately the control of the efficiency of each of its components.

4.2.1. Description of signals The HARPO trigger system is based on a PARISROC2 [19] chip mounted on a PMM2 [18] board. This trigger is built from the discriminated signals of the scintillators, laser and micromegas mesh as shown in Fig. 6. The available signals are:

- Laser: when a pulsed laser is used, the laser's trigger signal L (Fig. 5) is used. It also determines the time at which the event takes place, t_0, with a precision of a few 10 ns.

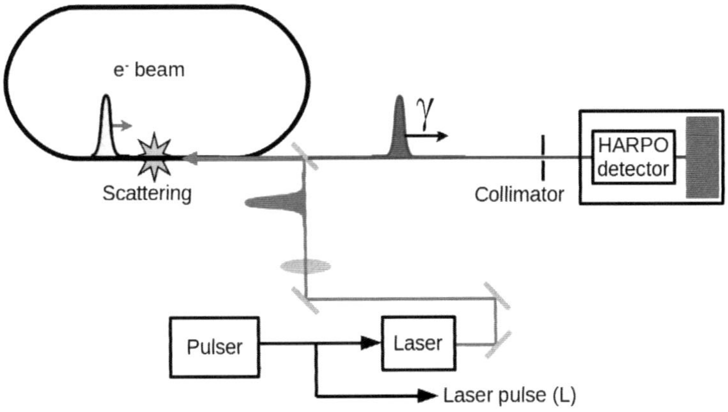

Figure 5. Schema of the Laser trigger used at NewSUBARU.

- Scintillator: the signal comes from 12 PMTs on 6 scintillators. The signal is recorded, whether there is a trigger or not. It gives t_0 with a hundred nanosecond precision. This signal determines t_0 when the laser signal is not available (CO_2).
- Micromegas mesh: the signal induced on the mesh is long and has an unpredictable shape: it corresponds to the time distribution of the charge deposited by the event in the TPC gas as it drifts towards the anode. The signal is amplified and derived through a 5 nF capacitor. It is then discriminated with a constant-fraction discriminator (CFD), so as to determine the rising edge t_{start} of the signal. The RC constant of the readout electronics of the mesh signal is $\approx 1\mu s$. It has been fixed as a compromise between the amount of electronic noise

Journal of Physics: Conference Series **650** (2015) 012016 doi:10.1088/1742-6596/650/1/012016

given the large capacitance of the full mesh, of 8 nF, and the need of a precise determination of the rising edge of the signal t_{start}.
Events that have a too small of a delay between the time of the event, t_0, and the discriminated mesh signal rising edge, t_{start}, are rejected. This provides a veto on background tracks created in the beam line upstream of the TPC and on gamma rays which converted in the material between the up scintillator and the active TPC gas (example: in the PCB supporting the amplification system).

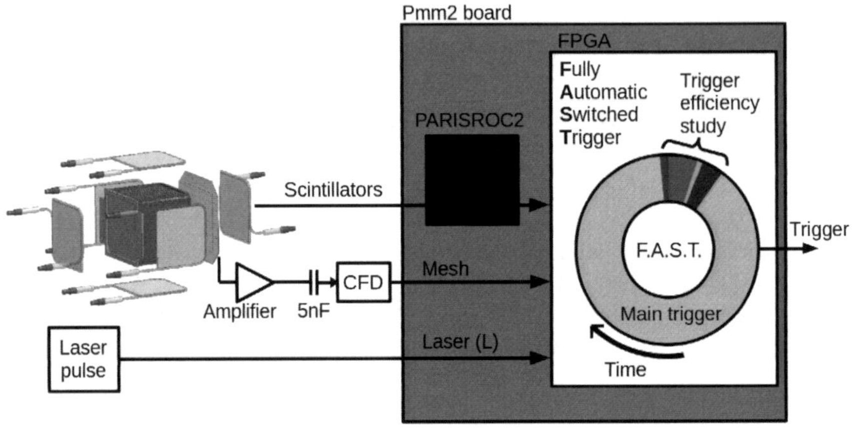

Figure 6. Global view of HARPO trigger system.

4.2.2. Gamma trigger line. The gamma trigger, T_γ, is designed so as to select gamma rays produced by LCS and that converted inside the TPC gas. The conditions are listed below:

- the upstream scintillator signal (S_{up}) is used as a veto,
- at least one downstream scintillator signal (O, other than "up") is required,
- the laser trigger signal (L), when present, is used,
- the events with a signal present in the mesh are selected, vetoing the prompt part as already explained, we require $t_{\text{start}} - t_0 > 1\,\mu\text{s}$. This "slow" selection of a mesh signal is denoted M_{slow}.

The main γ trigger line is therefore defined as the combination of the four following components:

$$T_\gamma \equiv \overline{S}_{up} \cap O \cap M_{\text{slow}} \cap L \tag{1}$$

The distribution of t_{start} is shown in Fig. 8. Gamma conversions inside the TPC gas (blue) are the signal events. The flatness of that part of the spectrum is due to the fact that the probability of conversion per unit path length is constant for a thin detector. Tracks entering the detector from upstream and gamma conversions in the detector material upstream of the TPC gas, which escaped the vetos, form the (green) peak. If we had used a trigger without any veto, the height of this peak would have been larger by several orders of magnitude, precluding an efficient data taking. Events (red) which lie outside the normal time range (i.e., $t_{\text{start}} < 100$ or $t_{\text{start}} > 400$) are due to energy deposition inside the gas by events that are not related to the trigger (pile-up).

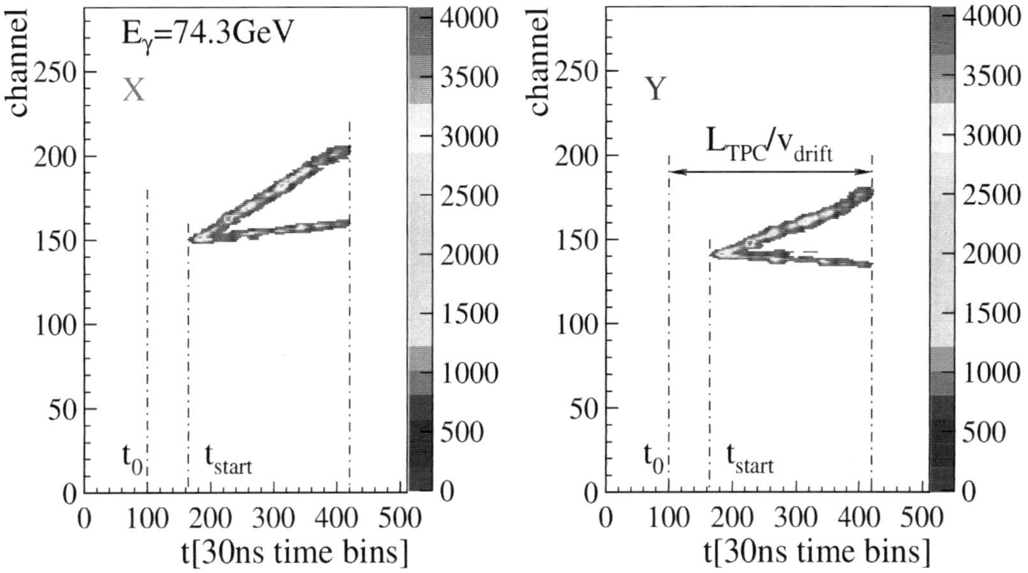

Figure 7. A gamma event that was selected by the gamma trigger, with definitions of timing information.

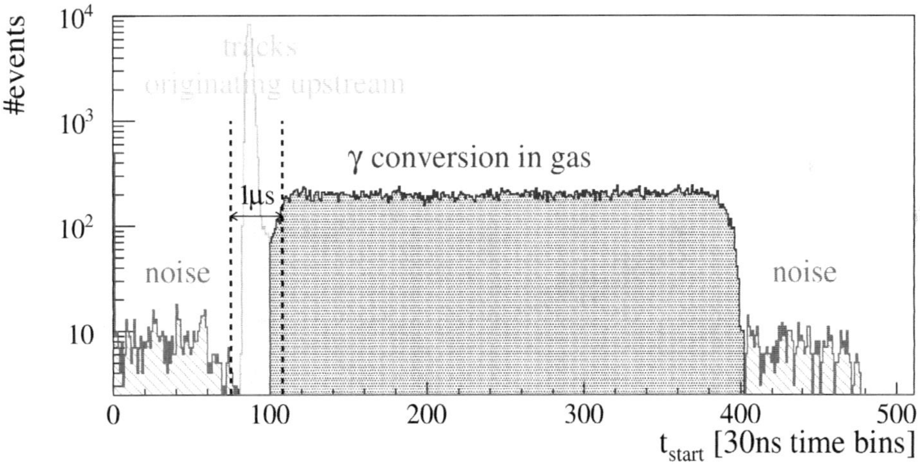

Figure 8. Distribution of event rising edge $t_{\rm start}$ for one run (all trigger lines): tracks originating upstream (green), gamma conversion inside the TPC gas (blue), charge deposition inside the gas by events not related to the trigger (red).

4.2.3. Other Trigger Lines. In addition to the main gamma trigger line, separate trigger lines have been formed with one of the trigger components omitted (such as $O \cap M_{\rm slow} \cap L$, $\overline{S}_{up} \cap M_{\rm slow} \cap L$, $\overline{S}_{up} \cap O \cap L$, $\overline{S}_{up} \cap O \cap M_{\rm slow}$, in the case of a pulsed laser). Analysis of these specific data will enable us to control separately the signal efficiency and the background rejection factor for each component of the main trigger line. Other dedicated lines were used, such as the "traversing z track" line for calibration purposes. Most of these additional lines were down-scaled so as to not saturate the bandwidth of the digitizing electronics, and were active only for a fraction of the time, as denoted by the rainbow part of Fig. 6. Approximately 90% of the time was devoted to the main trigger line.

4.3. Gas

The pressure vessel was evacuated down to $\sim 2 \times 10^{-6}$ bar, rinsed with ~ 0.1 bar gas, evacuated again, and then filled with 2.1 bar of the gas mixture (Ar:95 % Isobutane:5 %). This same gas was used in a sealed mode for 23 days.

At the end of the data taking, a pressure scan from 1 to 4 bar was performed, with the gas amplification tuned so that the signal amplitude was kept constant ($E_e = 1.5\,\mathrm{GeV}$, Nd 1ω, $P = 1$).

4.4. Monitoring

During data taking in the beam, some basic parameters were monitored, such as alignment, amplification gain, trigger performance and event quality after some very basic tracking.

4.4.1. Alignment. The alignment of the HARPO detector in the collimated gamma beam was monitored by plotting the transverse (x or y) position of the gamma conversion candidate vertex as a function of its longitudinal position z (Fig. 9). The vertex is defined here from the charge-weighted barycenter of the three first clusters of a selected gamma-conversion event (triggered by the gamma trigger).

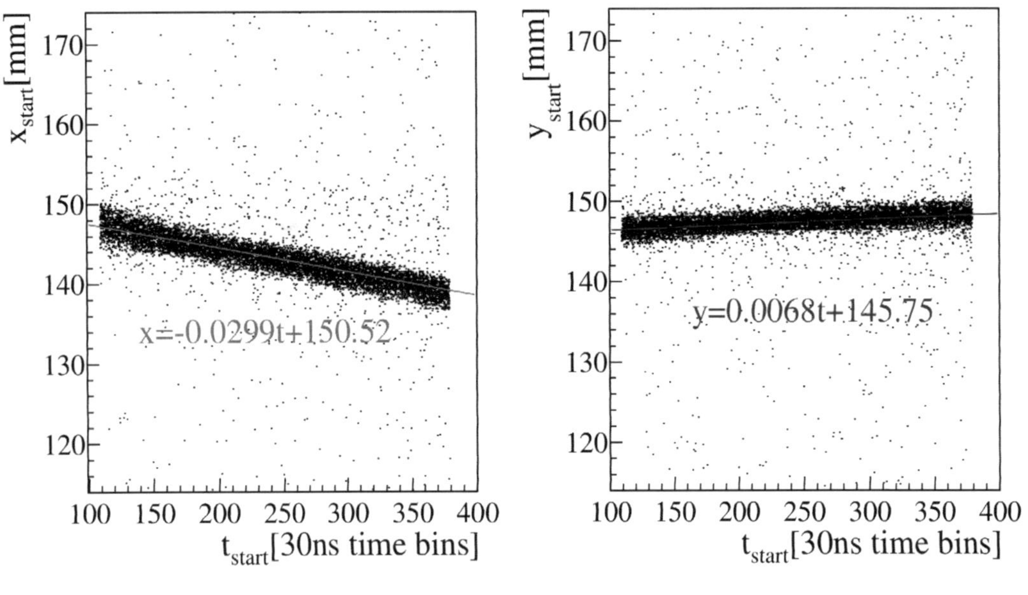

Figure 9. Alignment plot for run 1277 (1 mm diameter collimator), with linear fit (gamma trigger line).

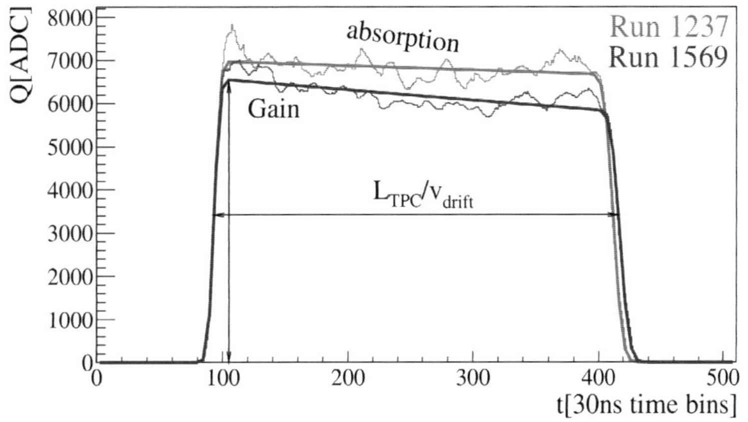

Figure 10. The average energy deposited at each time bin per traversing track along the z-axis for two runs separated by a full week of high flux running.

4.4.2. Calibration. The calibration of the TPC, i.e., the determination of the drift velocity, of the amplification gain and of the electron absorption, is performed by the analysis of the time distribution of the average charge per track as was described in section 3.2. A special trigger line ($\sim 2\%$ of the events) was used to select the tracks traversing in the z direction (i.e., along the drift time) without biasing the time distribution.

Fig. 10 and Fig. 11 show that there was no significant gas deterioration over the two weeks in the beam. The small features that can be seen in the time charts (Fig. 11) are associated with changes in the gamma-beam energy and/or of the laser pulsing rate, meaning that these small biases are related to the measurement, not to the quantity measured. The gas gain and the drift velocity were found to be sufficiently stable and the electron absorption due to either leaks or outgassing was found to be sufficiently small that we didn't have to renew the gas.

5. Conclusions and perspectives

A gas TPC is the instrument suited to covering the performance gap in cosmic gamma-ray detection in the MeV-GeV range. We have built a demonstrator that includes a micromegas+GEM amplification system. We have performed an experimental campaign with this demonstrator using the high-flux, quasi-monochromatic, polarised photon beam at NewSUBARU. Tests that were performed during data taking indicate an excellent detector performance.

We have more than 1 TB of data ($> 6 \times 10^7$ events, a large fraction of which is estimated to be gamma conversions in the gas) to analyze so that we can study the gamma conversion to e^+e^- pairs and, in particular, measure the low energy polarization asymmetry, and characterize the performance of the demonstrator both as a gamma telescope and as a gamma polarimeter. Further development is ongoing to meet the constraints of a space environment [14], in particular, tests on the behaviour of the detector in space with a dedicated trigger. The next step will then be to verify its operation with a balloon flight.

6. Acknowledgments

This work is funded by the P2IO LabEx (ANR-10-LABX-0038) in the framework "Investissements d'Avenir" (ANR-11-IDEX-0003-01) managed by the French National Research Agency (ANR), and directly by the ANR (ANR-13-BS05-0002).

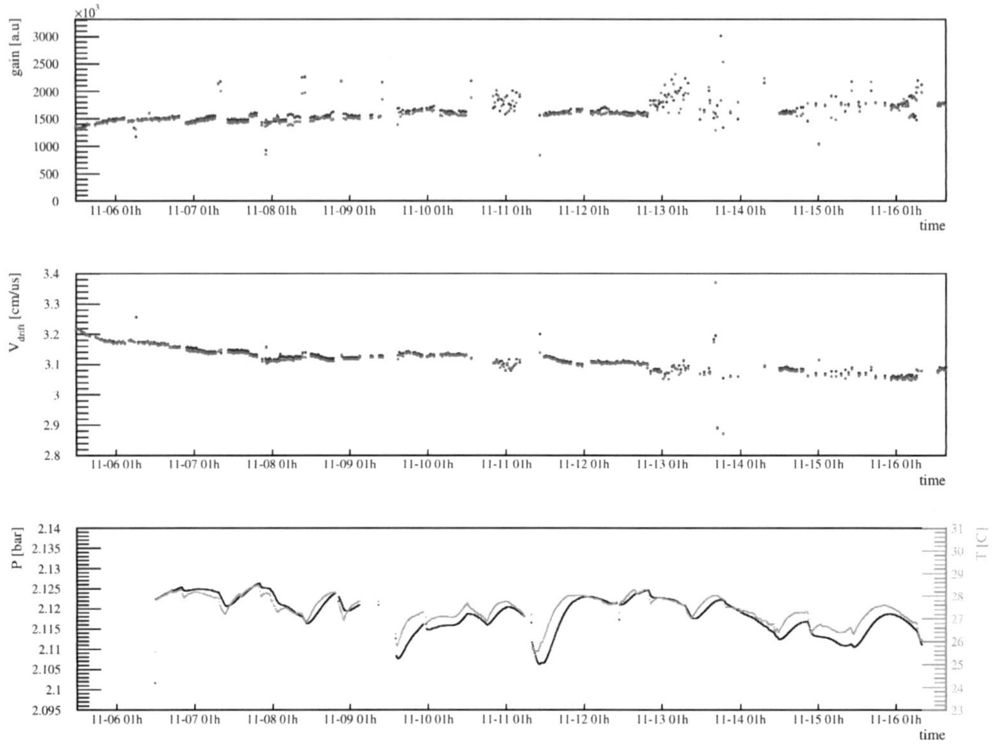

Figure 11. Time variation of the gain, estimated by the average charge per track, corrected for the track angle that is normalized to 30 cm track length, the drift velocity (x in red, y in blue) and the pressure (black) and temperature (green) of the gas.

This work was performed by using NewSUBARU-GACKO (Gamma Collaboration Hutch of Konan University).

References

[1] D. Eichler, A. Levinson, *Polarization of GRB via scattering off a relativistic sheath*, Astrophys. J. **596**, L147 (2003). [astro-ph/0306360].

[2] H. Zhang and M. Böttcher, *X-Ray and Gamma-Ray Polarization in Leptonic and Hadronic Jet Models of Blazars*, Astrophys. J. **774**, 18 (2013) [arXiv:1307.4187 [astro-ph.HE]].

[3] M. Forot, P. Laurent, I. A. Grenier, C. Gouiffes and F. Lebrun, "Polarization of the Crab pulsar and nebula as observed by the Integral/IBIS telescope," Astrophys. J. **688**, L29 (2008) [arXiv:0809.1292 [astro-ph]].

[4] H. Bloemen, "Diffuse galactic gamma-ray emission," Ann. Rev. Astron. Astrophys. **27** (1989) 469.

[5] M. Ackermann *et al.* [Fermi-LAT Collaboration], *The Fermi Large Area Telescope On Orbit: Event Classification, Instrument Response Functions, and Calibration*, Astrophys. J. Suppl. **203**, 4 (2012) [arXiv:1206.1896 [astro-ph.IM]].

[6] D. Bernard, *TPC in gamma-ray astronomy above pair-creation threshold*, Nucl. Instrum. Meth. A **701**, 225 (2013) [Erratum-ibid. A **713**, 76 (2013)] [arXiv:1211.1534 [astro-ph.IM]].

[7] D. Bernard, *Polarimetry of cosmic gamma-ray sources above e^+e^- pair creation threshold*, Nucl. Instrum. Meth. A **729**, 765 (2013) [arXiv:1307.3892 [astro-ph.IM]].

[8] Mattox, J. R., Mayer-Hasselwander, H. A., Strong, A. W., *Analysis of the COS B data for evidence of linear polarization of VELA pulsar gamma rays*, Astrophys. J. 363 (1990) 270.

[9] D. Bernard, *HARPO - A Gaseous TPC for High Angular Resolution Gamma-Ray Astronomy and Polarimetry from the MeV to the TeV*, Nucl. Instrum. Meth. A **718**, 395 (2013). [arXiv:1210.4399 [astro-ph.IM]].

[10] F. Sauli, *GEM: A new concept for electron amplification in gas detectors*, Nucl. Instrum. Meth. A **386**, 531 (1997).
[11] Y. Giomataris, P. Rebourgeard, J. P. Robert and G. Charpak, *MICROMEGAS: A high-granularity position-sensitive gaseous detector for high particle-flux environments*, Nucl. Instrum. Meth. A **376**, 29 (1996).
[12] P. Baron, D. Calvet, E. Delagnes *et al.*, *AFTER, an ASIC for the readout of the large T2K time projection chambers*, IEEE Trans. Nucl. Sci. **55**, 1744 (2008).
[13] D. Calvet, "A Versatile Readout System for Small to Medium Scale Gaseous and Silicon Detectors", IEEE Transactions on Nuclear Science 61 (2014) 675.
[14] D. Bernard *et al.*, Presented at SPIE2014. Proc SPIE 9144 M. [arXiv:1406.4830 [astro-ph.IM]].
[15] S. Amano *et al.*, *Several-MeV γ-ray generation at NewSUBARU by laser Compton backscattering*, Nucl. Instrum. Meth. A **602** 2009 337.
[16] Ph. Gros, *HARPO - TPC for High Energy Astrophysics and Polarimetry from the MeV to the TeV*, TIPP 14 3rd Technology and Instrumentation in Particle Physics conference, 2-6 June 2014 Amsterdam, Proceedings PoS(TIPP2014)133.
[17] M. Weisskopf, *X-ray Polarimetry: from the early days to an outlook for the future*, X-ray polarisation in astrophysics, August 25-28, Stockholm, Sweden.
[18] B. Genolini *et al.*, *PMm2: Large photomultipliers and innovative electronics for the next-generation neutrino experiments*, Nucl. Instrum. Meth. A **610** (2009) 249 [arXiv:0811.2681 [physics.ins-det]].
[19] S. Conforti Di Lorenzo *et al.*, *PARISROC, an autonomous front-end ASIC for triggerless acquisition in next generation neutrino experiments*, Nucl. Instrum. Meth. A **695** (2012) 373.

Institute of Physics
Dirac House, Temple Back
Bristol BS1 6BE UK

ISSN: 1742-6588
ISBN 978-1-5108-1813-2

9 781510 818132